OFFICIAL COPY.

104
Gen. No.
3265

3397
40
W.O.
2227

MUSKETRY REGULATIONS.

PART I.

1909.

(REPRINTED, WITH AMENDMENTS, 1914.)

GENERAL STAFF, WAR OFFICE.

LONDON:
PRINTED UNDER THE AUTHORITY OF HIS MAJESTY'S STATIONERY OFFICE
BY HARRISON AND SONS, 45-47, ST. MARTIN'S LANE, W.C.,
PRINTERS IN ORDINARY TO HIS MAJESTY.

To be purchased, either directly or through any Bookseller, from
WYMAN AND SONS, LTD., 29, BREAMS BUILDINGS, FETTER LANE, E.C., and
54, ST. MARY STREET CARDIFF; or
H.M. STATIONERY OFFICE (SCOTTISH BRANCH), 23, FORTH STREET, EDINBURGH; or
E. PONSONBY, LTD., 116, GRAFTON STREET, DUBLIN;
or from the Agencies in the British Colonies and Dependencies,
the United States of America, the Continent of Europe and Abroad of
T. FISHER UNWIN, LONDON, W.C.

Price Sixpence.

These Regulations are issued by command of the Army Council for the guidance of all concerned.

R. H. Brade

War Office,

26th September, 1914.

(B 11476) Wt. 20878—132 200M 8/15 H & S P. 14/350

CONTENTS.

SEC. PAGE

List of Plates xi

CHAPTER I.

1. General instructions 1

CHAPTER II.

ARMS AND AMMUNITION.

DESCRIPTION OF ARMS.

2. The Short Magazine Lee-Enfield Rifle, Marks III and IV 5
3. Action of the mechanism, &c. 12
4. Miniature rifles and aiming tubes 16
5. The Charger Loading Magazine Lee-Enfield Rifle, Mark 1* 23
6. (Cancelled) ... ... —
7. The Webley Pistol 27
8. Action of the mechanism, &c. 28

CARE OF ARMS.

9. General instructions 30
10. Instructions for cleaning 35
11. Instructions for care of arms and ammunition ... 40
12. Examination of small arms 42
13. Testing arms... 46

AMMUNITION.

SEC.		PAGE
14.	S.A.A., Marks VI and VII	50
15.	Allowance of ammunition	52
16.	Defective ammunition	54

CHAPTER III.

THE THEORY OF RIFLE FIRE AND ITS PRACTICAL APPLICATION.

17.	Application of theory	56
18.	Definitions	56
19.	Rifling	59
20.	Forces acting on the bullet	59
21.	Elevation	60
22.	Sighting of rifles	60
23.	Jump	61
24.	Drift	62
25.	Effect of fixing the bayonet, of resting the rifle, and of a heated or oily barrel...	62
26.	Dangerous space	64
27.	Ricochets—Trajectory and Range Tables ...	66
28.	Firing up and down hill	72
29.	Barometric pressure and temperature	73
30.	Wind and light	74
31.	Use of collective fire	75
32.	The dispersion of individual and collective fire ...	76
33.	Ground in relation to fire effect	78
34.	Searching	81
35.	Night firing	82

Chapter IV.

MUSKETRY EXERCISES.

AIMING INSTRUCTION.

SEC.		PAGE
36.	Stages of instruction	84
37.	Accuracy in aiming	85
38.	Common faults in aiming	87
39.	Triangle of error	88
40.	Aiming at the ground, and marking down an enemy	90
41.	Aiming off for wind	90
42.	Aiming up and down	92
43.	Rapid adjustment of sights	92
44.	Aiming off for movement	93

FIRING INSTRUCTION.

45.	Instructors	94
46.	Firing rest	95
47.	Trigger pressing	95
48.	Firing standing	97
49.	Firing prone	101
50.	Firing in other positions	102
51.	Use of cover	103
52.	Rapid loading	103
53.	Use of the safety catch and cut-off	104
54.	Muscle exercises	105
55.	Fire discipline	106
56.	Collective fire	107
57.	Rates of fire	108
58.	Description of targets	109

SEC. PAGE
59. Words of command... 110
60. Preliminary exercises 111
61. Fire in two ranks 113
62. Webley pistol exercise 113
63. Tests of elementary training 116
64. Method of conducting the tests 117

CHAPTER V.

VISUAL TRAINING AND RANGING.

65. General 122
66. Visual training 123
67. Judging distance 125
68. Tests in judging distance 127
69. Range finding by observation of fire 129
70. Test of range-finding instruments... 130
71. Training and test of range-takers 130

CHAPTER VI.

PRELIMINARY TRAINING AND RANGE PRACTICES.

PRELIMINARY TRAINING.

72. Preliminary training of recruits 138
73. Preliminary training of trained soldiers 139
74. Miniature cartridge practice 140
75. Thirty yards ranges... 143

INSTRUCTIONAL RANGE PRACTICES.

76. General 144
77. Grouping 145

SEC. PAGE
78. Application 147
79. Special instructions 148
80. Recruits' course, Regular Forces, Cavalry, R.E. and Infantry 151
81. Execution of Tables A and B in the same year ... 151

TRAINED SOLDIERS' COURSES, REGULAR FORCES.

82. General 152
83. Special instructions 153
84. Allotment of ammunition and computation of averages 155
85. Conditions of qualification 156
86. Classification practices and conditions of classification 157
87. Royal Engineers 158
87A. R.A.M.C. and A.V.C. 159

THE CONDUCT OF RANGE PRACTICES.

88. General instructions 159
89. Special instructions 161
90. Webley pistol practice 166
91. Range duties... 168
92. Signalling and scoring 175

CHAPTER VII.

FIELD PRACTICES.

93. General 182
94. Special instructions 185
95. Individual field practices 187
96. Fire direction practices 189

SEC. PAGE
97. Collective field practices 191
98. Tactical schemes. Section and half-company exercises 193
99. Standard tests of fire effect 194
100. Comparative tests of fire effect and vulnerability... 195
101. Company exercises. Combined field firing ... 197

Chapter VIII.

COMPETITIONS, BADGES, PRIZES, AND METAL FUNDS.

102. Competitions... 200
103. Badges 204
104. Judging distance badges 208
105. Prizes... 209
106. Metal funds 210

Chapter IX.

REPORTS, RETURNS AND RECORDS.

107. General 212
108. Annual returns 212
109. Registers 218
110. Diaries 220
111. Territorial Force returns 221
112. Machine gun returns 222

Chapter X.

MACHINE GUNS.

113. General 223
114. Machine gun course... 232
115. (Omitted) ... ... —

CHAPTER XI.

SCHOOL OF MUSKETRY.

SEC.		PAGE
116.	Object of the school	249
117.	The commandant	250
118.	Courses of instruction	250
119.	Record of qualifications	252
120.	Serjeant-instructors	253

PAGE

APPENDIX I.

Cavalry, Royal Engineers, and Infantry, Tables A and B 254

APPENDIX II.

Royal Artillery, Army Service Corps, and Army Ordnance Corps 262

APPENDIX III.

Special Reserve, Tables A and B 269

APPENDIX IV.

Territorial Force, Tables A and B 274

APPENDIX V

Channel Islands Militia, Tables A and B 283

APPENDIX VI

Officers Training Corps 290

APPENDIX VII.

HAND GRENADE (MARK I).

SEC.

1. General description 291
2. To prepare the grenade for use 293
3. To throw the grenade 294
4. Instructions for use 295

INDEX 297

LIST OF PLATES.

			PAGE
I.	Nomenclature, Plates II to IV	... *Facing*	4
II. III. IV.	Short Rifle, Magazine Lee-Enfield (Mark III) ...	*Following*	4
V.	Magazine Platform (S.M.L.-E.)	... ...	15
VI.	Nomenclature, Plates VII to X	... ...	22
VII.	Charger Loading, Lee-Enfield	*Follywing*	22
VIII.	" " " "		
IX.	" " " "		
X.	" " " "		
XI.	Diagram showing dangerous space (Marks VI and VII, S.A.A.)	...	64
XII.	Diagram showing dispersion of fire ...	...	76
XIII.	Cone of fire	*Following*	76
XIV.	Rectangles of shot groups ...		
XV.	Diagram showing the depth dispersion of concentrated collective fire		
XVI.	Diagram showing the zone of effective fire	*Facing*	77
XVII.	Ground in respect of fire (1) ...	... ...	79
XVIII.	" " " (2) ...	... ...	80

Plates XIX—XXXIII following p. 104.

XIX. Standing, position when loading.
XX. „ „ „ firing.
XXI. „ „ „ „
XXII. Lying, position when firing.
XXIII. „ „ „ „
XXIV. Kneeling, position when firing.
XXV. „ „ „ „
XXVI. Lying behind isolated cover.
XXVII. Kneeling behind cover, position when loading.
XXVIII. „ „ „ „ „ loaded.
XXIX. „ „ „ „ „ firing.
XXX. „ „ „ „ „ „
XXXI. Sitting, position when loaded.
XXXII. „ „ „ firing.
XXXIII. „ „ „ „

XXXIV. Hand grenade, Mark I ... *facing page* 291

NOTE.—Amendments, other than minor corrections, are indicated by a black line placed in the margin.

MUSKETRY REGULATIONS,

PART I, 1909.

(REPRINTED, WITH AMENDMENTS, 1914.)

CHAPTER I.

Section 1.—*General instructions.*

1. The purpose of musketry training is to render the individual soldier proficient in the use of small arms, to make him acquainted with the capabilities of the weapon with which he is armed, and to give him confidence in its power and accuracy. Musketry training should also qualify officers and non-commissioned officers to direct and control fire under service conditions, and should provide commanders with information as to the power of rifle-calibre weapons, and the abilities of those who use them. It should therefore assist commanders, when they have considered the conditions under which fire is delivered in war, in making the most effective use of their commands in the presence of an enemy.

2. Musketry must not be regarded as a form of skill-at-arms which can be separated from tactical study and manœuvre training. It is to be associated in theory and practice with manœuvre.

3. Musketry training must vary according to the range facilities and ground available; it will generally include thorough preliminary training of recruits and trained soldiers, range practices, field practices, and theoretical instruction of fire leaders.

4. Field practices should be regarded as the most important of these, but ample time must always be allotted to preliminary training.

5. Field practices are divided into (*a*) Individual, and (*b*) Collective; Instructional range practices should always precede Individual Field practices at home stations, at stations abroad the sequence of practices may be modified by the general officer commanding if he so desires; Collective Field practices may be fired at any time of the year.

6. These regulations apply to all arms of the Regular Army, Special Reserve, and Territorial Force. In the case of the Special Reserve and Territorial Force certain variations are dealt with in supplementary instructions included in the regulations issued for those branches.

7. The words *battalion* and *company* throughout these regulations will be read, where necessary, to mean *regiment, squadron*, or *battery*.

8. The musketry year at Home stations for the Regular Forces and Special Reserve and Channel Islands Militia will begin on the 1st January. At stations abroad, permanent dates have been fixed to meet climatic conditions, as follows:

North China ...	1st January.
Egypt	
Bermuda	
Straits Settlements*	
Gibraltar	

* In the case of a British infantry battalion transferred from South China, the beginning of the next musketry year may be deferred to the 1st April following the arrival at the new station.

Ceylon	...	1st February.
South China	...	1st April.
Jamaica	...	1st April.
Mauritius	...	1st January.
West Africa	...	1st October.
Malta	...	1st October.
South Africa	...	16th November.

This paragraph will not apply to India.

The musketry year for the Territorial Force will begin on the 1st November.

9. A soldier is termed a *recruit* until he has completed Table A, when he is called a *trained soldier.*

10. All commanders will ensure that the musketry training of units under their command, armed with rifle or carbine, is carried out in accordance with the spirit of these regulations. Officers commanding battalions and regiments are allotted a special musketry staff to assist in the training of recruits and in other musketry duties. Company commanders are to be given full responsibility for the musketry efficiency of the trained soldiers of their commands.

11. General officers commanding-in-chief will correspond direct with the commandant, School of Musketry, Hythe, on musketry questions.

PLATE I.

Names of the Parts of Rifle, Short M.L.E., Mark III, Referred to in Plates II to IV.

1. Blade foresight.
2. Foresight block.
3. Band foresight block.
4. Key ,, ,,
5. Crosspin ,, ,,
5A. Backsight bed.
6. ,, ,, crosspin.
6A. ,, ,, sight spring screw.
7. Backsight leaf.
8. ,, slide.
9. ,, slide catch.
10. ,, fine adjustment worm wheel.
10A. Windgauge.
10B. ,, screw.
11. Backsight ramps.
12. Seating for safety catch.
13. Safety catch.
14. Locking bolt stem.
15. Bolt.
16. ,, head.
17. Striker.
18. Cocking-piece.
19. Striker collar with stud.
20. Bolt head tenon.
21. Cocking-piece locking recesses.
22. Locking bolt.
23. ,, ,, flat.
24. ,, ,, thumb-piece.
25. ,, ,, aperture sight stem.
26. Locking bolt stop pin recesses.
27. ,, ,, safety catch stem.
28. ,, ,, ,, ,, arm.
29. ,, ,, screw threads.
30. ,, ,, seating.
31. Bolt cam grooves.
32. Sear.
33. ,, seating.
34. ,, spring.
35. Magazine catch.
36. Full bent of cocking-piece.
37. Short arm of sear.
38. } Trigger ribs.
39. }
40. Trigger.
41. Trigger axis pin.
41A. Magazine case.
41B. ,, platform spring.
41C. ,, auxiliary ,,
42. Guard trigger.
43. Stock fore-end.
44. Spring and stud fore-end.
45. Protector backsight.
46. Handguard front and rear.
47. Spring handguard gear.
48. Lower band groove.
49. Lower band.
50. Nosecap.
51. Protector foresight.
52. Sword bar.
53. Boss for ring of sword bayonet crosspiece.
54. Swivel seating.
55. ,, piling.
56. Nosecap barrel opening.
57. Inner band.
58. ,, ,, screw.
59. ,, ,, ,, spring.
60. Butt sling swivel.
61. Sword bayonet, pattern '07.
62. Bridge charger guide.
63. Cut-off.

PLATE II.

SHORT RIFLE, MAGAZINE

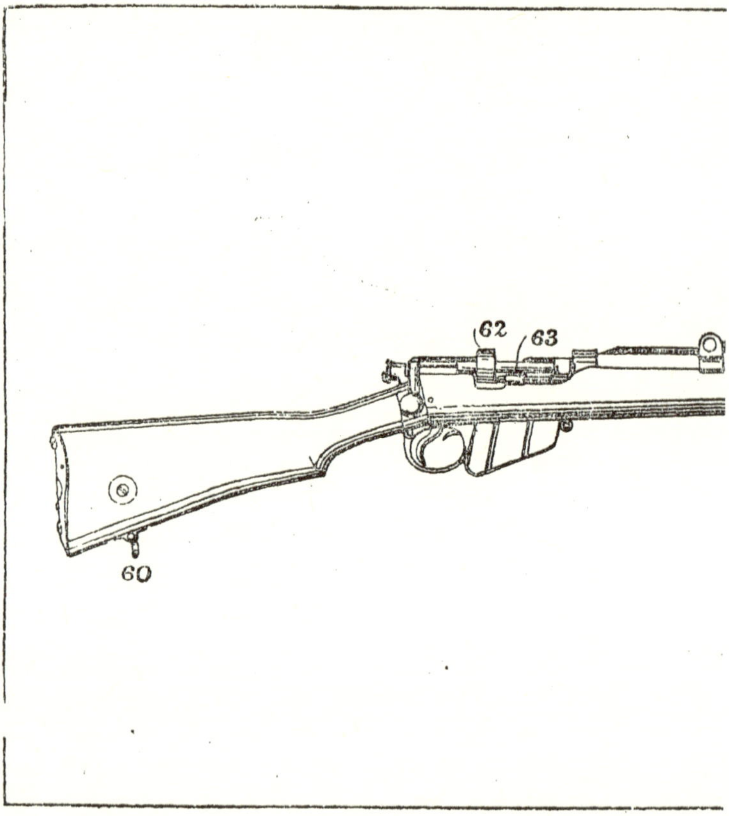

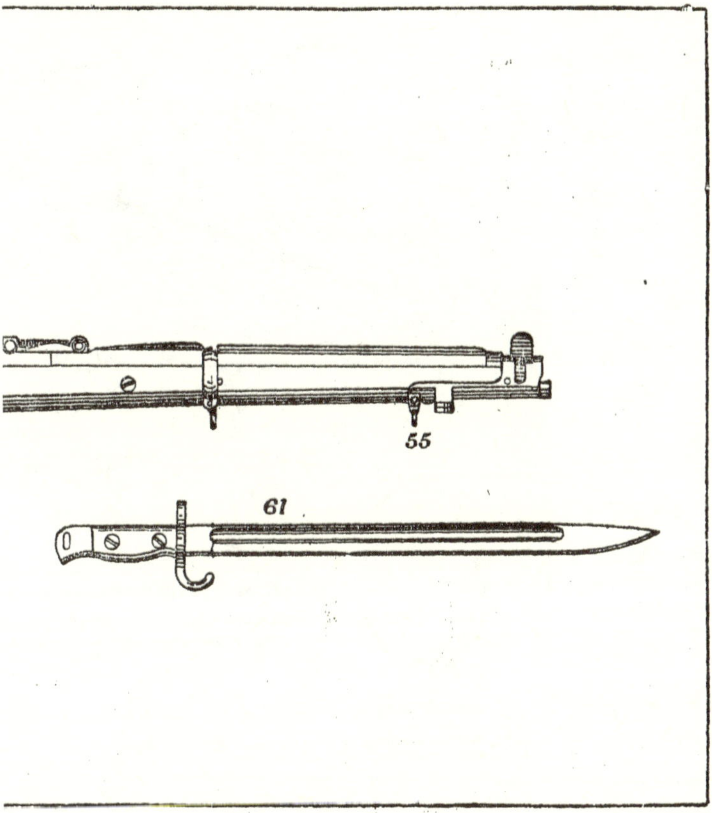
55
61

PLATE III.

SHORT RIFLE, MAGAZINE LEE-ENFIELD (MARK III).

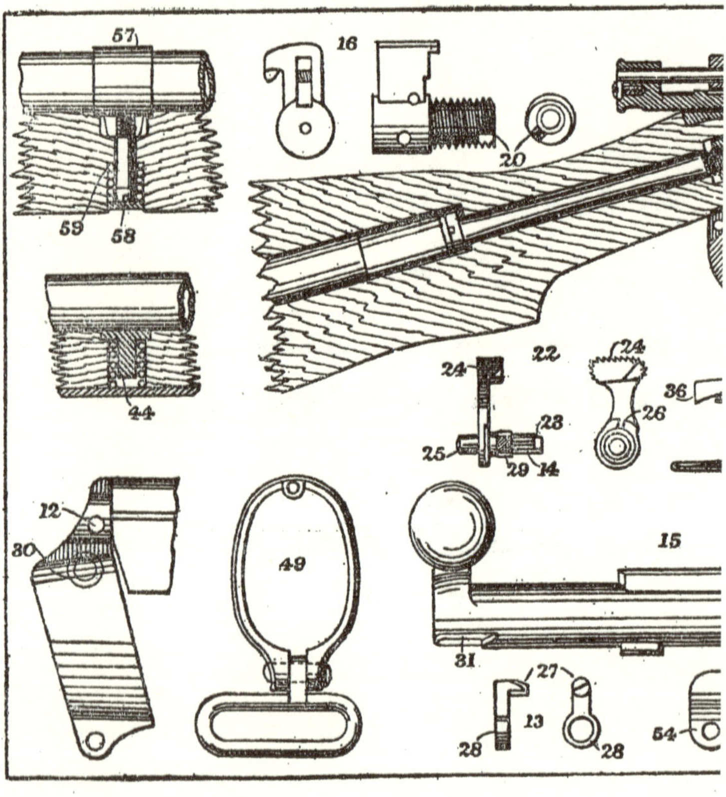

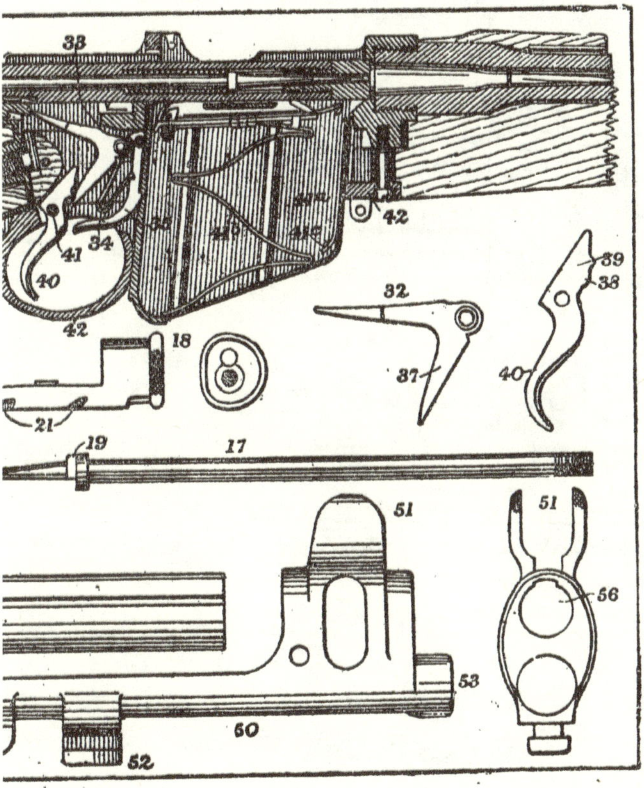

33
42
41
34
40
42
18
21
32
37
39
38
40
19
17
51
51
56
53
50
52

PLATE IV.

SHORT RIFLE, MAGAZINE LEE-ENFIELD (MARK III).

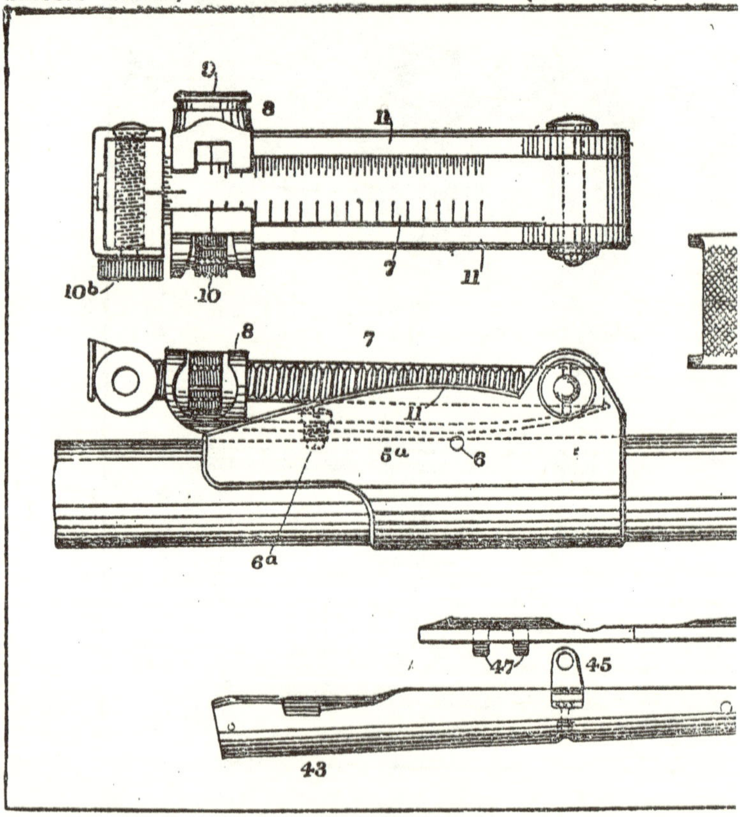

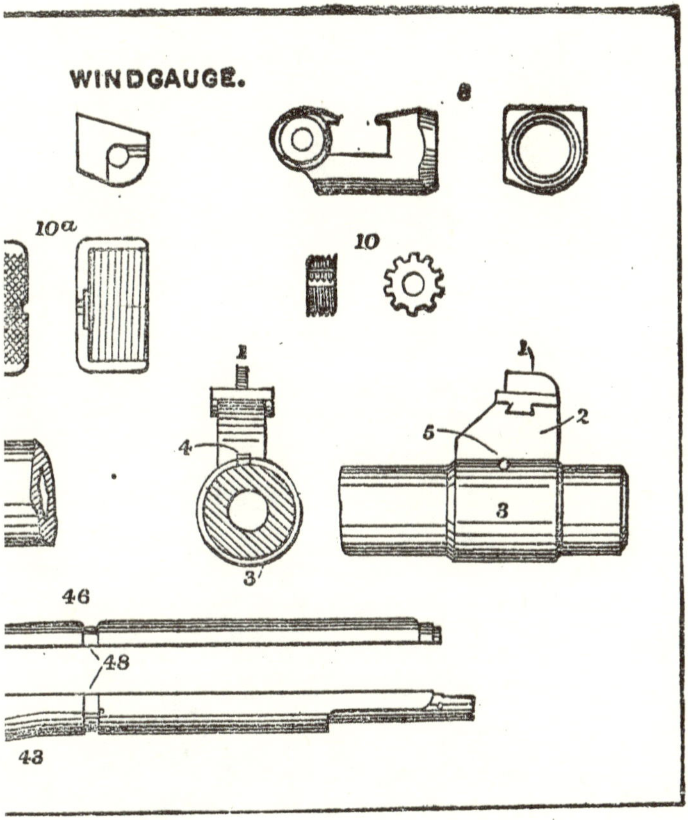

WINDGAUGE.
8
10a
10
1
1
4
5
2
3
3
46
48
43

CHAPTER II.

ARMS AND AMMUNITION.*

DESCRIPTION OF ARMS.

Section 2.—The Short Magazine Lee-Enfield Rifle, Marks III and IV.

12. [Information regarding points in which Marks I, I*, II, and II* differ from Marks III and IV, is given in square brackets.]

Weight of rifle with magazine empty, Mark III, about	8 lbs. $10\frac{1}{2}$ ozs.
Weight of rifle with magazine empty, Mark IV, about	8 lbs. $14\frac{1}{2}$ ozs.
[Weight of rifle with magazine empty, about	8 lbs. $2\frac{1}{2}$ ozs.
Weight of sword-bayonet without scabbard, about	1 lb. $\frac{1}{2}$ oz.
Length of butt, long, about	1 ft. $1\frac{13}{16}$ ins.
" " normal, about ...	1 ft. $1\frac{5}{16}$ ins.
" " short, about	1 ft. $0\frac{13}{16}$ ins.
" rifle with normal length of butt, without sword bayonet, about	3 ft. $8\frac{1}{2}$ ins.
" rifle with normal length of butt, with sword bayonet, about	5 ft. 2 ins.
" barrel, about	2 ft. $1\frac{1}{4}$ ins.
Calibre	·303 in.
Rifling system	Enfield.

* NOTE.—For further information *see* "Text Book of Small Arms," 1909

Twist of rifling ...	... left hand; 1 turn in 10 ins. or 33 calibres.	
Muzzle velocity	... Mark VI ammunition about 2,060 f.s. Mark VII ammunition, about 2,440 f.s.	
Number of grooves	... ...	5.
Depth of grooves	at muzzle	·0065 in.
	at breech, to within 14 ins. of muzzle	·00575 in.
Width of lands ...	... ...	·0936 in.
Sighting system.	Adjustable blade foresight, radial backsight, with vertical adjustment and windgauge.	
Method of loading	... Charger, holding 5 cartridges.	
[Sighting system.	Adjustable barleycorn foresight, radial backsight, with vertical adjustment and windgauge.]	

13. The Short Magazine Lee-Enfield rifle is constructed on the bolt system, the breech being closed by a bolt worked by a lever on the right side. The bolt contains the mainspring and striker.

14. The striker is inserted from the front of the bolt; its rear end passes through the mainspring, and screws into the cocking-piece, thus connecting the cocking-piece and bolt. The front of the mainspring bears against the collar on the striker, and the rear end against the back of the bolt chamber.

15. The bolt-head is a separate component. It carries the extractor, extractor screw, and a V-shaped spring, and is connected with the bolt by means of a screwed tenon, in which is cut a slot which acts as a key for removing the striker from the cocking-piece. A hook on the right of the bolt-head, which engages with a rib on the body, secures the bolt-head when the bolt is rotated.

[The head has a slide for the bolt-head charger guide.]

16. On the left of the body, about a cartridge length from the mouth of the chamber, the end of the ejector screw projects into the bolt way. In front of the ejector screw, the left side of the body is cut away to afford clearance for the thumb, when loading with the charger.

17. Immediately in rear of the thumb clearance is the bridge charger guide.

[There is also a raised portion in rear of the thumb clearance, in which a slot is cut. The guides into which the charger is inserted when loading are formed by the slot cut in the raised portion of body behind the thumb clearance and the slot in the bolt-head charger guide.]

18. A cut-off which consists of a flat piece of steel hinged to the right side of the body is provided and is moved by means of a thumb piece.

19. On the left of the body a socket is arranged to receive the safety catch, which consists of a small transverse locking bolt and pin. The former is threaded at the part where the pin works. On the thumb piece being forced back, it engages in one of the recesses cut in the cocking-piece, which is then locked. The pin is brought into action by means of the screw on the locking-bolt; its front end engages in the short cam groove at the rear end of the bolt, and thus prevents the rotation of the bolt.

20. There are two holes or gas escapes in the body, one on either side of the breech, to facilitate a lateral escape of gas in case of a burst cartridge; there is also a third hole or gas escape in the bolt-head, in case of a blow-back of gas from a defective cap.

21. The magazine is made from sheet steel, and holds 10 cartridges. The cartridges rest on a platform which is actuated by a spring made from ribbon steel. To afford a smooth bearing for the platform an auxiliary spring is fixed by a lip and stud to the front of the magazine. There are four patterns of magazines—Nos. 1 and 2 for use with Mark VI ammunition, and Nos. 3 and 4 for use with Mark VI or Mark VII ammunition. Nos. 1 and 2 magazines are fitted on the right side with a stop clip and on the left side with a fixed lip to retain the platform in its place and keep the top cartridge in position. For these purposes No. 3 magazine has a stop clip on the right side and a spring lip on the left side, and No. 4 magazine is provided with fixed lips on either side. The magazine can be released by pressing the catch which projects through the trigger guard in front of the trigger.

22. The upper band and nose-cap are formed in one piece, with an extension in front, on which the ring of the cross-piece of the bayonet fits, and a bar underneath the rear end, to hold the pommel of the bayonet. Lugs are provided to carry a swivel if required. Wings are attached to protect the foresight. The outer band carries the band swivel. The inner band, by which also the barrel is attached to the fore-end, is one inch in rear of the outer band. It is fixed by a screw passing through the fore-end, round which is a spring to maintain the necessary pressure against the barrel.

[The inner band is immediately under the outer band.]

23. The barrel is supported by the fore-end, except from about ½-inch in rear of the inner band recess to the knox-form.

[The fore-end is freed from the barrel throughout, except for about ½-inch in front and ½-inch in rear of the inner band, and under the knox-form at the breech-end.]

24. The butt and fore-end are separate. The butt has a butt trap cover with spring to enable an oil bottle and pull-through to be carried in the stock bolt hole. In the butt four holes are bored longitudinally for lightness, it has a pistol grip, and a brass butt plate fixed by two screws; on the right side is a brass disc (Mark II* and IV, strap on butt plate), on which the regiment, number, and the date of issue are marked. In front is the butt swivel stem to which the swivel is attached. Butts are issued in three lengths; one ½ an inch longer, and one ½ an inch shorter than the normal, marked respectively on the heel L and S; the normal bears no mark. (*See* para. **119.**) The rear end of the body is formed with a socket into which the butt fits, and is held in position by the stock bolt, which is squared at the front end to fit the keeper plate in the fore-end, thus preventing the stock bolt from turning, and the butt from becoming loose. When stripping the rifle the fore-end must first be removed before turning the stock bolt.

25. The barrel is screwed into the body and secured to the fore-end by the inner band, nose-cap, outer band, and trigger guard.

26. A wooden handguard completely covers the barrel. It is in two pieces, being divided at the centre of the back-sight bed; the front portion is held by the outer band, and by a metal cap, which fits into a recess between the nose-cap and the barrel; the rear portion is held by springs gripping the barrel. Both portions rest on the fore-end, and are quite free of the barrel.

[The backsight protector is attached to the rear handguard.]

27. The foresight is of blade pattern. For Mark VI ammunition it is made in five heights (marked -015, 0, 015, 03 and 045 respectively) and for Mark VII ammunition in seven heights marked -06, -045, -03, -015, 0, 015, and 03 respectively) to allow any deviation from the normal in the shooting of individual rifles to be corrected. It is adjustable laterally. It is dovetailed into the foresight block band, which is secured to the barrel by means of a key and pin.

[The foresight is of barleycorn pattern and for Mark VI ammunition is made in five heights (marked ·93, ·945, ·96, ·975 and ·99 respectively). For Mark VII ammunition it is made in six heights (marked ·9, ·915, ·93, ·945, ·96 and ·975 respectively).]

28. The backsight is fitted with a leaf and has a curved ramp on each side of the bed on which the slide rests and by means of which elevation is obtained.

29. The backsight protector is level with the rear end of the leaf, and is fixed to the fore-end with a screw and nut.

30. The leaf is pivoted to the front end of the bed, and is graduated on the right by lines for every hundred yards from 200 to 2,000 yards, the even numbers only being marked by figures.

31. The slide can be adjusted for every 50 yards, and is retained in position by the worm wheel, which engages in a rack on the right of the leaf. The wheel is disengaged from the rack by pressing the stud on the left inwards. The lines on the left of the leaf give graduations for 25 yards.

[The slide can be adjusted for every 50 yards, and is retained in position by means of one of two catches engaging in a notch on the side of the leaf. These catches are disengaged from the notches by pressing the bone studs on each side of the slide.

[A vertical fine adjustment and a windgauge are fitted to the leaf.]

32. Small increases and decreases of elevation can be obtained by rotating the worm wheel, the periphery of which is divided by thumb-nail notches into 10 parts, each part giving a difference of 5 yards of range. One complete revolution of the wheel thus corresponds to a difference of 50 yards.

33. The windgauge is adjusted by means of a screw with milled head on the right. It is provided with a scale, each division on which represents 6 inches per 100 yards. Between the screw head and the windgauge is a small friction spring which prevents accidental movement of the windgauge. This spring gives a click at each quarter revolution of the screw, and 6 clicks, or one and a half revolutions of the milled head, are equivalent to one division on the scale.

[The fine adjustment is joined to the leaf by a dovetail, and can be used for obtaining any additional elevation required between the 50 yards elevations given on the leaf. A scale is provided on the left edge of the fine adjustment and leaf. The windgauge is fitted in the vertical fine adjustment and is moved by a screw and provided with a scale. A thumb-nail slot is cut in the top front edge of the fine adjustment as a guide for centering the windgauge. Each division on the two scales represents 6 inches per 100 yards.]

34. On the left side of the rifle long range sights are provided for ranges from 1,600 to 2,800 yards. These sights consist of an aperture sight pivoted to the rear of the body, and a dial sight, with pointer, secured by means of a screw to the fore-end in rear of the outer band. The ranges are marked on the dial plate.

Section 3.—Action of the mechanism, &c.

35. On raising the bolt lever the bolt is rotated to the left, thereby forcing the stud on the cocking-piece to move backward from the long to the short groove in the rear of the bolt; this action withdraws the striker about $\frac{1}{8}$th of an inch. At the same time, a steel lug on the under side of the bolt works down an inclined slot on the left side of the body, withdrawing the bolt about $\frac{1}{8}$th of an inch, and effecting primary extraction.

[On drawing back the bolt, the charger guide on the bolt head, coming in contact with the top on the top right side of the body, is held stationary in position to receive the charger.]

36. The charger containing 5 cartridges is placed between the guides, and the cartridges are forced into the magazine by the thumb.

37. On pushing the bolt forward, the charger is thrown out and the full bent of the cocking-piece is brought against the nose of the sear. The cocking-piece and striker are thus held stationary whilst the bolt travels forward, the mainspring being compressed between the collar of the striker and the rear end of the mainspring chamber in the bolt.

38. During the forward movement, the lower part of the bolt head engages behind the upper part of the base of the top cartridge in the magazine and pushes the cartridge into the chamber.

39. On turning the bolt to the right, the breech is finally closed by the rib on the bolt working over the resisting shoulder on the right side of the body; at the same time the lug on the bolt works into the recess cut on the left side of the body.

40. On pressing the trigger, when the rifle is at full cock, the two ribs on the trigger bear in succession on the lower arm of the sear and produce a double-pull off, the first pull bringing the nose of the sear to the bottom of the bent of the cocking-piece, and the second pull finally releasing the cocking-piece; the mainspring then carries the striker forward, exploding the charge.

41. The shock of the discharge is taken equally on either side of the body; on the right, by the bolt rib bearing against the resisting shoulder, and on the left, by the bolt lug bearing against the rear wall of the recess in the body.

42. If the bolt has not been properly turned over when the trigger is pressed, the stud on the cocking-piece either causes the breech to close automatically by striking against the rounded corner of the division between the two grooves of the bolt, causing the bolt to turn down and the breech to close, or else the stud on the cocking-piece strikes full against the division between the two grooves, and prevents the striker flying forward. If then the bolt is closed by hand, the whole action becomes locked as the sear nose is engaged by the half bent, which is undercut, whilst the cocking-piece stud travels halfway down the longer groove: consequently the trigger cannot be pressed, nor can the bolt be rotated until the action is placed at full cock by drawing back the cocking-piece.

43. On opening and drawing back the bolt, the cartridge case is drawn out of the chamber by the extractor, and is ejected when its base comes in contact with the ejector screw.

44. The safety-catch may be used when the cocking-piece is either at full-cock or in the fired position. In neither case can the cocking-piece be moved backward or forward, nor the bolt be rotated. Care should be taken to see that the bolt lever is in its lowest position before moving the safety catch, as if it is slightly raised when the safety catch is moved there is a possibility of the aperture sight spring being strained and weakened.

45. *To strip the bolt.*—Remove the extractor spring, extractor screw, and extractor. Remove the striker keeper screw, and see that the stud on cocking-piece is in the long cam.

Unscrew the bolt head, and remove the striker, mainspring, and cocking-piece. If the striker is so tight in the cocking-piece that the bolt-head cannot be moved by hand, place the head in the bolt hole of the body and turn the bolt about a quarter of a turn over to the left. If it is found that the striker is not released by unscrewing the bolt-head to the full extent, draw back the cocking-piece until the stud is clear of the end of the bolt and turn it to the right, so that the stud rests on the rear end of the bolt; screw home the bolt-head; then replace the cocking-piece, with the stud in the long cam, and unscrew the bolt-head.

[Remove the charger guide stop screw and charger guide. Remove the extractor spring, extractor screw and extractor. Raise the striker keeper nut, if fitted, out of the locking position, turn it slightly, or remove the striker keeper screw, and see that the stud on the cocking-piece is in the long cam. Remove the striker as in Mark III. Replace the striker keeper nut in locking position, and unscrew keeper nut screw, and remove screw, spring, and nut.]

46. *To assemble the bolt.*—Place the mainspring, striker, and bolt-head in the bolt, and screw home the bolt-head for about six turns; then place the cocking-piece on the bolt, **so that the stud is in the long cam,** and screw home the bolt-head; draw back the cocking-piece until the stud is clear of the end of the bolt and turn it until the stud rests on the rear end of the bolt. Unscrew the bolt-head about six turns; turn the cocking-piece so that the stud moves into the long cam. Screw home the striker until the end is flush with the rear end of the

To face page 15.

Plate V.

※ No. 3 MAGAZINE.

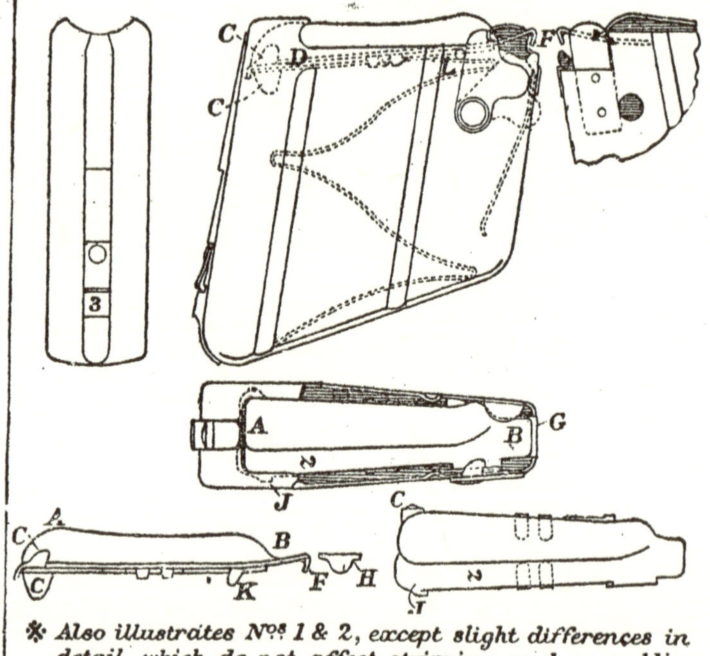

※ *Also illustrates Nos. 1 & 2, except slight differences in detail, which do not affect stripping and assembling.*

No 4 MAGAZINE.

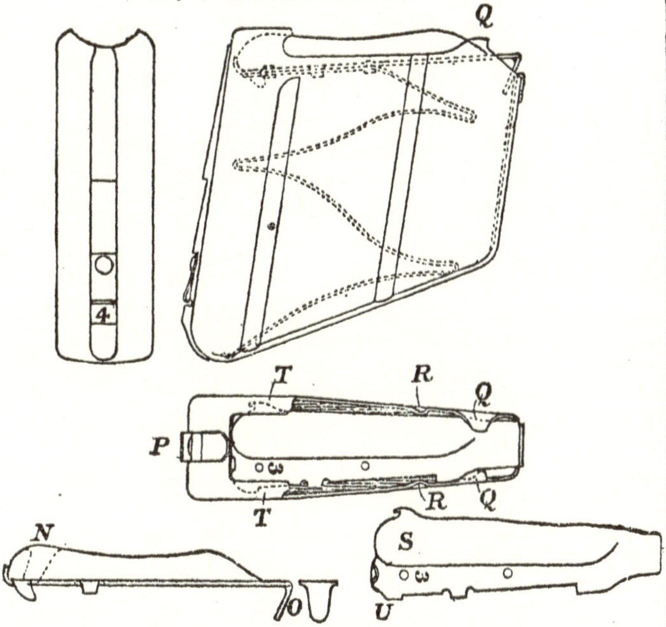

cocking-piece, and the keeper-screw recess in the striker is in correct position; replace the keeper-screw, move the cocking-piece to the short cam and screw home the bolt-head, being careful to see that it is screwed fully home before replacing the bolt in the rifle. If it is found impossible to screw home the striker in the cocking-piece by hand, extra leverage may be obtained by placing the bolt-head in the bolt hole of the body. Replace the extractor, extractor screw, and extractor spring.

[Replace in the cocking-piece the keeper-nut, spring, and screw (if fitted) and lift the nut out of the locking position. Replace the mainspring, striker, and bolt-head as for Mark III, and screw the striker home till the keeper-nut (or screw) recess is in correct position. Place the keeper-nut in the locking position (or replace the keeper-screw). Screw home the bolt-head as for Mark III. Replace extractor, extractor screw, extractor spring, charger guide and stop screw.]

47. *To remove the magazine platform.*—Nos. 1, 2, and 3 magazines, push the stop clip forward to the disengaged position shown in dotted lines, slightly depress the rear end of the platform at A (Plate V), and lift the front end at B with the fingers, at the same time pressing the platform to the right side and upwards; then pull the platform forward with a slight wriggling lateral movement (taking care to keep the rear side-ears C above the ribs of the case D) and lift the platform with its spring out of the magazine. Under no circumstances must leverage be applied to the lip of the platform at the front end F to force it out. If force is found necessary a knife blade may be placed on top of the auxiliary spring at the front of the magazine at G, with the point resting under the front of the platform at the left side of the lip at H and a slight upward pressure will release the platform.

No. 4 magazine, depress the rear end of the platform at N (Plate V) as far as possible, at the same time holding up the front end; then pull the front end at O towards the rear end of the case P until it passes under the front lips Q Q and between the inner forward ribs R R in case through which it is sprung. The front end of the platform should then rise up out of the case. Then tilt the rear end of the platform N sideways—the left side S uppermost and draw forward out of the case.

48. *To assemble the magazine platform.*—Nos. 1, 2 and 3 magazines, place the spring in the case and, tilting the right side of the platform down slightly at J, press back the rear end of platform into the case just above the ribs D; then see that the front ears K are just in front of the ribs of the case at L, slightly depress the rear end of the platform at A, and press the front end down with a sharp pressure at M; then push back the stop clip against the stop.

No. 4 magazine, insert the rear end N of the platform in front of the rear lips T T on the case, tilting it sideways so that the right side U enters first; then depress the rear end until the front end is below the level of the front lips Q Q on the case, then press forward, forcing the front end through the internal ribs R R in the case.

Section 4.—Miniature rifles and aiming tubes.
(a) The Short Rifle, ·22 in. Rim-fire, Mark I.

49. Weight of rifle, about	8 lbs. 4 ozs.
Length of rifle, about	3 ft. 8½ ins.
Length of barrel, about	2 ft. 1¼ ins.
Calibre	·214 in.
Rifling system	Segmental.

Twist of rifling	Right hand, 1 turn in 16 ins.
Number of grooves	8.
Depth of grooves (mean)	·0052 in.
Width of lands	·03 in.

Sighting system :—Adjustable blade foresight, radial backsight, with fine adjustment and windgauge as on the Short Magazine, Lee-Enfield, Mark III.

The rifle is bored and chambered to take the rim-fire aiming tube cartridge, Mark I, but is made to approximate as nearly as possible to the Short, Magazine, Lee-Enfield Rifle, Mark III, in length and weight, from which, however, it differs in the following respects :—

i. The striker hole in the face of the bolt-head is bored eccentric to the axis of the bolt-head. The bolt-head is marked " ·22 " on the hook.

ii. The striker is arranged suitably for the bolt-head, being flattened on the front portion to clear the rear end of the extractor. (With the strengthened pattern of bolt-head the striker is in two parts—the firing pin portion being separate from the main portion.)

iii. An extractor of suitable form, having a long and narrow hook, is fitted. It is so arranged that the outward movement is limited by the body to prevent injury to the extractor spring in the event of a burst cartridge case.

iv. The cut-off is not fitted.

v. The gas escapes on the right and left of the body are enlarged.

vi. The magazine is not fitted. The bottom of the body is left open to allow empty cases to fall out.

vii. The nose cap is similar to that of the long rifle, but has no fittings for the bayonet. The upper band passes round it and round the barrel.

viii. The fore-end is shorter than in the short rifle.

ix. The front hand guard is not fitted.

x. A foresight protector is fitted on the barrel and fixed by a screw.

xi. The long range sights are not fitted.

xii. The rifles are sighted before issue so that the 300 yards graduation on the backsight gives the elevation required for use at 25 yards with the ·22 rim-fire cartridge.

50. *To strip the bolt.*—Remove the striker keeper-screw. Draw back the cocking-piece and turn it until the stud rests on the rear end of the bolt. Remove the bolt-head. Turn the cocking-piece and allow it to fall into the short cam. Grip the front end of the striker in a vice which has suitable protecting clams, care being taken not to damage the striker or bend its eccentric point, and unscrew the cocking-piece, using the bolt handle as a lever. Remove striker, mainspring, and cocking-piece.

To assemble the bolt.—Place the cocking-piece in the long cam and insert the mainspring and striker in the bolt. Using a vice as in stripping, screw home the cocking-piece. Replace the striker keeper-screw. Draw back the cocking-piece, and turn it till the stud rests on the rear end of the bolt. Screw home the bolt-head and place the cocking-piece in the short cam.

The bolt-head must not be used for removing or replacing the striker.

*(b) Aiming tubes for Short and Long Lee-Enfield Rifles.**

51. The tubes are of steel, rifled with eight flat-bottomed grooves. As they are adapted for use with the rim-fire cartridge they can only be used with the special bolt in which the striker is eccentric.

The instructions to be followed in fitting the tubes are as follows:—

Remove the bolt as described for the short M.L.E. rifle; unscrew the nut at the muzzle end of the tube and remove the gun-metal and leather washers. Insert the tube in the barrel from the breech end, placing it in such a position that the projection on the end of the sliding extractor coincides with the extractor way in the barrel. Replace the washers and screw the nut up tightly by hand. Replace the bolt.

If the tube is fixed correctly in the barrel the extractor on the bolthead should leave the sliding extractor of the tube after withdrawing it about $\frac{7}{16}$ inch from the face of the chamber. Neglect to fix the tube in the proper position may render it unserviceable.

To remove the tube from the barrel, reverse these operations.

The bolt will be stripped in the same manner as the bolt of the short rifle, ·22-inch R.F. (para. **50**).

(c) The War Office Miniature Rifle.

52. The rifle is on the bolt principle, and is about 4 inches shorter and considerably lighter than the Service short rifle. It is bored and chambered to take the rim-fire aiming tube cartridge, Mark I.

The rifling is segmental, of the Metford type.

---

* For instructions for the employment of aiming tubes, *see* para. **364.**

Though the bolt differs from that of the short rifle, ·22-inch R.F., previously described, the action of the mechanism is much the same.

A magazine is fitted which holds 5 cartridges. In charging the magazine the base of the cartridge is pressed down on the magazine platform and the point of the bullet then allowed to fall forward. Care must be taken that the base of the cartridge is held down by the shoulders at the rear end of the magazine.

The magazine is removed by pressing the catch, which will be found under the stock between the magazine and the trigger guard. The bottom of the magazine can be removed by sliding it to the rear, when the spring and platform can be taken out.

The foresight is of blade pattern and is fitted with a hinged sight protector, and the backsight has a U notch and a slide which rests on a curved ramp and which can be adjusted to give any elevation required. Lines are marked on the leaf showing the elevations necessary for 25, 50, 100, 150 and 200 yards. A windgauge is fitted in the cap of the backsight and is adjusted by milled head screws.

The bolt is removed for cleaning by pressing back the trigger as far as it will go and withdrawing the bolt.

PLATE VI.—NAMES OF THE PARTS OF RIFLES, CHARGER-LOADING M.L.E., REFERRED TO IN PLATES VII TO X.

1. Backsight bed.
2. ,, ,, ramp.
3. Aperture sight.
4. Dial sight plate.
5. ,, ,, pointer.
6. Aperture sight peep hole.
7. ,, ,, pivot screw
8. ,, ,, spring.
9. ,, ,, spring ribs.
10. ,, sight cross-cut notches.
11. Pivot aperture sight pointer.
12. ,, screw.
13. Spring disc.
14. Bead pointer.
15. Screwed boss, fixing dial sight.
16. Centring pin dial sight.
17. Index point of pointer.
18. Bolt rib.
19. Extractor seating.
20. Bolt lug.
21. Tongue of cocking-piece.
22. Recess for lug of bolt.
23. Body bead.
24. Retaining catch.
25. Resistance shoulder.
26. Cut-off slot.
27. Body socket.
28. ,, ,, boss.
29. Stock bolt.
30. Gas escape hole.
31. Bolt breech.
32. Lever bolt breech.
33. Lug seating.
34. Long cam groove.
35. Short ,, ,,
36. Cam-shaped face.
37. } 38. } Cam grooves separating studs.
39. Bolt head.
40. ,, ,, tenon.
41. ,, ,, projection.
42. Extractor slot.
43. Extractor.
44. ,, axis screw.
45. ,, spring.
46. ,, ,, pin.
47. ,, ,, ,, hole.
48. Bolt head hook.
49. Striker.
50. Cocking piece.
51. Striker collar.
51A. Cocking-piece full bent.
52. ,, ,, half ,,
54. ,, ,, seating for safety catch.
55. Safety catch.
56. Striker keeper-screw.
57. Recess in striker for the keeper-screw.
58. Safety catch stem.
59. Safety catch finger piece.
60. ,, ,, locking grooves.
61. ,, ,, plunger.
62. Seating for plunger and spiral spring of safety catch.
63. Nipple holes for plunger of safety catch.
64. Retaining catch spring.
65. Ejector.
66. Sear.
67. ,, screw.
68. ,, seating in body.
69. ,, spring.
70. Trigger.
71. ,, axis pin.
72. Magazine box or case.
73. } 74. } ,, platform guides.
75. ,, stop clip.
76. ,, ,, ,, stud.
77. ,, platform.
78. ,, clip stop.
79. ,, tooth.
80. ,, catch tooth.
81. ,, catch.
82. ,, ,, finger piece.
83. Magazine platform plate.
84. ,, ,, ,, tongues.
85. ,, ,, spring.
86. ,, ,, auxiliary spring.
87. Cut-off.
88. ,, screw.
89. ,, thumbpiece.
90. ,, catch.
91. Guard trigger.
92 Screw guard front.
93. ,, ,, collar.
94. ,, ,, ,, rear.
95. Lower band.
96. Sling swivel.
97. Nose-cap.
98. ,, band.
99. ,, fixing screw.
100. Piling swivel.
101. Diagonal slot in nose-cap to clear the fore-sight block when assembling, &c.
102. Stock bolt, plate keeper.
103. Stock butt swivel.
104. Oil bottle, Mark III (latest Mark IV).

PLATE VII.

CHARGER LOADING LEE-ENFIELD.

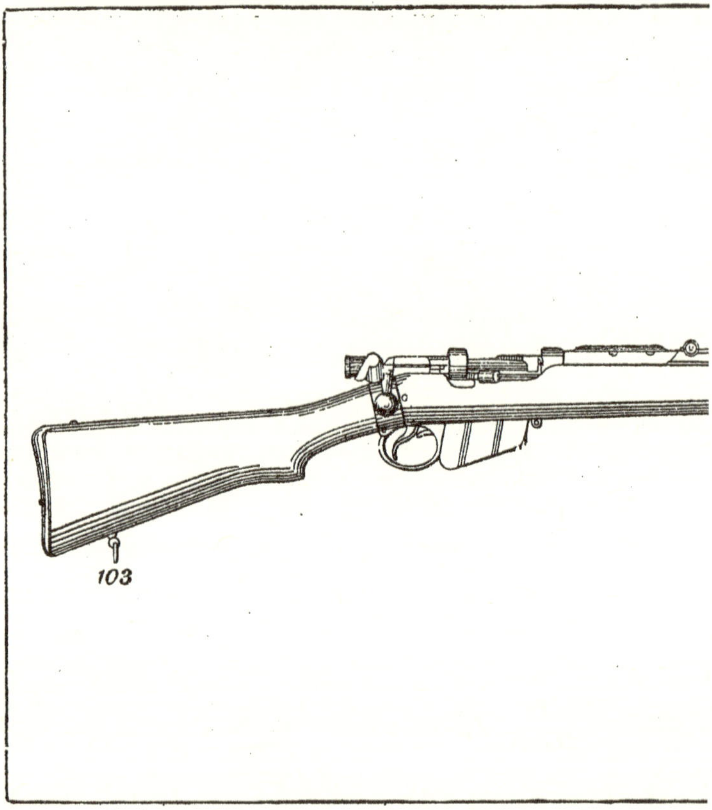

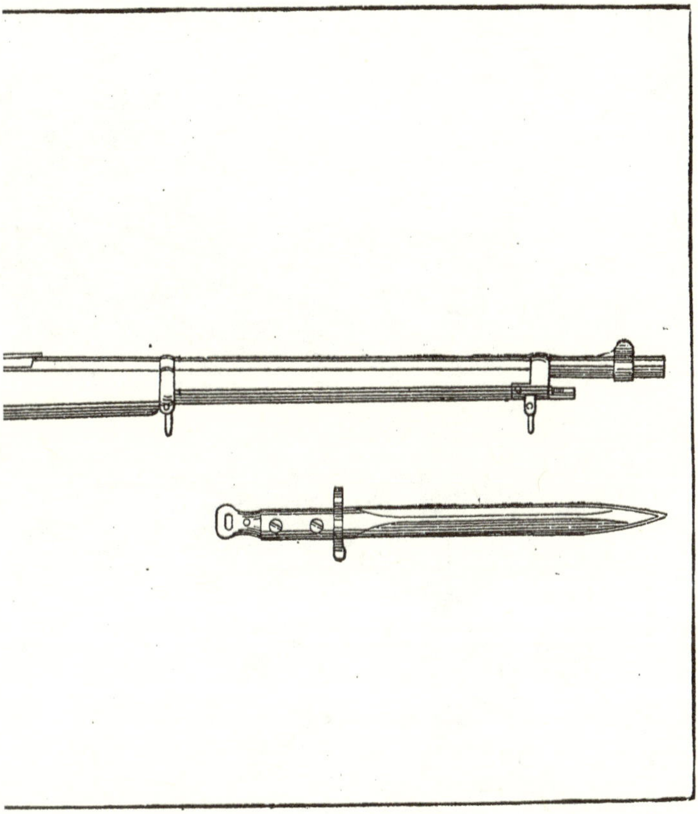

PLATE VIII.

CHARGER LOADING LEE-ENFIELD.

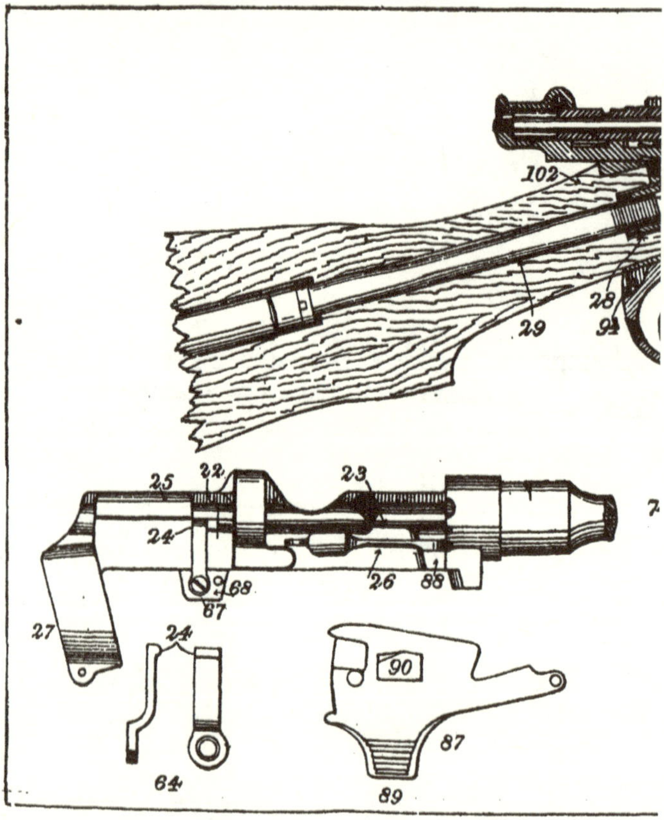

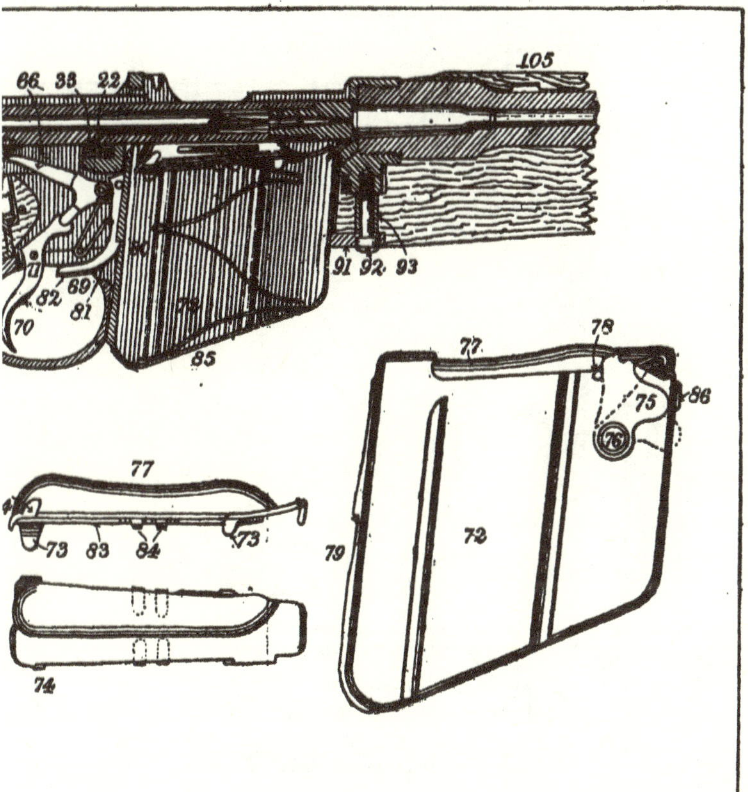
105
66
33
22
91
92
93
82
69
81
70
85
77
78
75
86
76
77
73
83
84
73
72
79
74

Plate IX.

CHARGER LOADING LEE-ENFIELD.

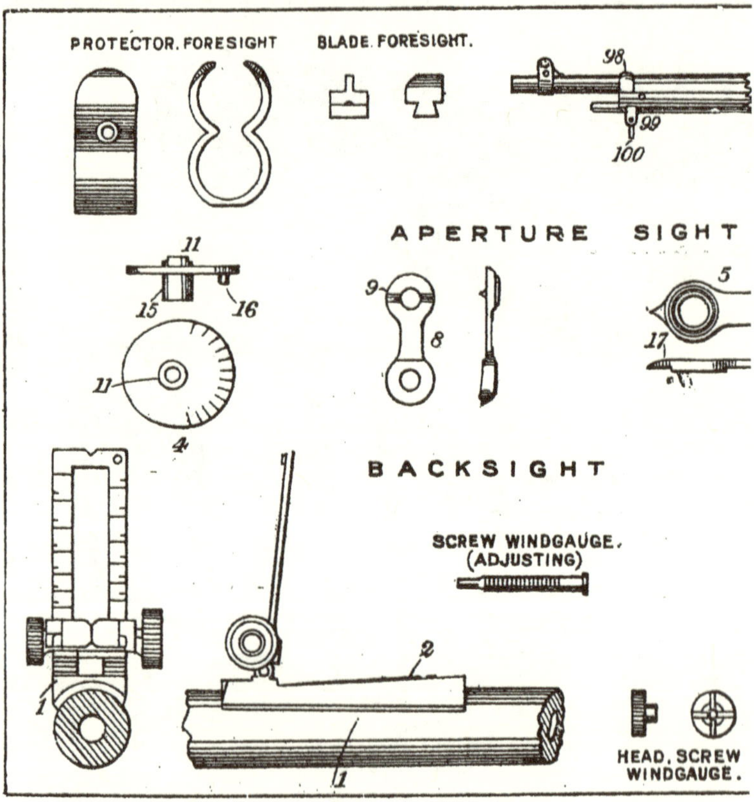

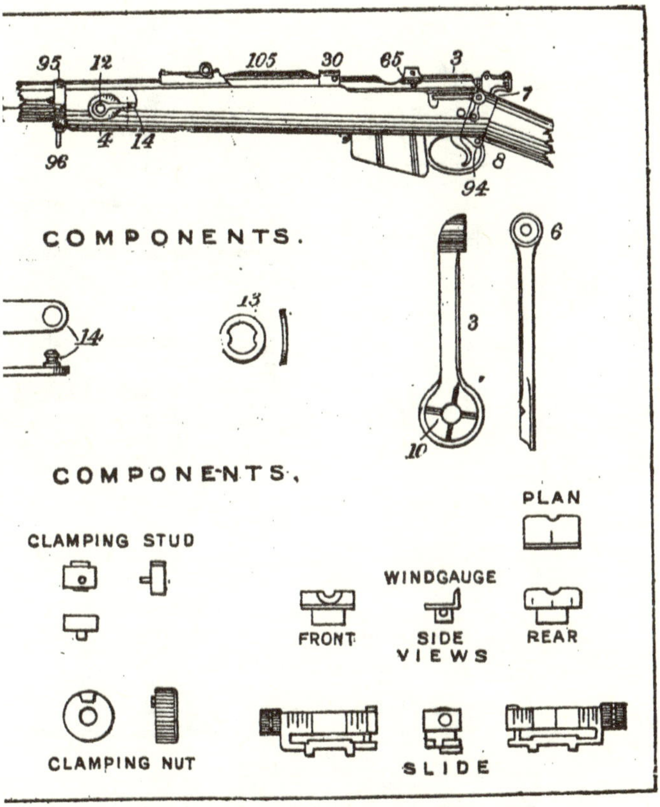
95
12
105
30
65
3
7
96
4
14
8
94
COMPONENTS.
6
13
14
3
10
COMPONENTS.
PLAN
CLAMPING STUD
WINDGAUGE
FRONT
SIDE
VIEWS
REAR
CLAMPING NUT
SLIDE

PLATE X.

CHARGER LOADING LEE-ENFIELD.

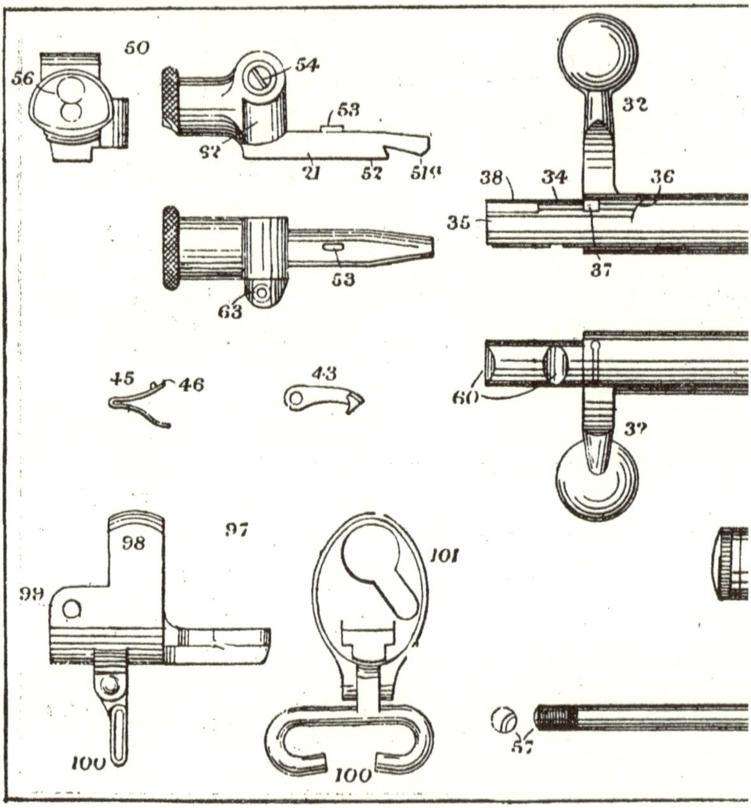

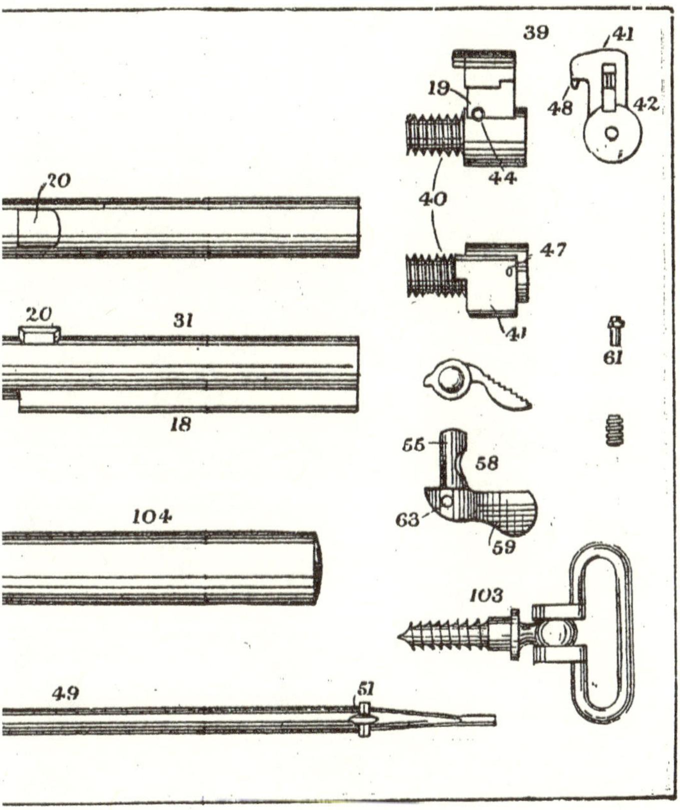
39
41
19
48
42
44
20
40
47
20
31
41
61
18
55
58
104
63
59
103
49
51

Section 5.—The Charger-loading Magazine Lee-Enfield Rifle, Mark I*.

53. Weight, including oil bottle and pull-through, about	9 lbs. 5 ozs.
Weight of sword bayonet without scabbard	15½ ozs.
Length of rifle without bayonet ...	4 ft. 1½ ins.
Length of rifle with bayonet ...	5 ft. 1½ ins.
" barrel	2 ft. 6 8/100 ins.
Calibre	·303 in.
Rifling system	Enfield.
Twist of rifling	Uniform, left-handed, 1 turn in 10 in. or 33 calibres.
Muzzle velocity (about) ...	Mark VI ammunition 2,060 f.s.
" "	Mark VII ammunition 2,440 f.s.,
Number of grooves	5.
Depth of grooves	·0065 in.
Width of lands	·0936 in.

54. The Lee-Enfield rifle is constructed on the bolt system, the breech being closed by a bolt worked by a lever on the right side. The bolt contains the mainspring and striker.

55. The striker is inserted from the front end of the bolt; its rear end passes through the mainspring and screws into the cocking-piece, thus connecting the cocking-piece and bolt. The front of the mainspring bears against the collar on the striker, and the rear end against the back of the bolt chamber.

56. A safety catch is fitted, consisting of a small transverse bolt in the cocking-piece, which is operated by a finger-piece on the right. On turning up the finger-piece the safety bolt engages in a slot in a projection on the rear end of the bolt; this locks the cocking-piece and bolt together.

57. The bolt-head is a separate component. It carries an extractor screw and V-shaped spring, and is connected with the bolt by means of a tenon and screw. A hook on the bolt-head which engages with a rib on the body secures the bolt-head when the bolt is rotated.

58. On the left-hand side of the body, about a cartridge length from the mouth of the chamber, the end of the ejector screw projects into the bolt-way.

59. In front of the ejector screw the left side of the body is cut away to afford clearance for the thumb when loading with the charger. In rear of this clearance is the bridge charger guide.

60. There are two holes or gas escapes in the body, one on either side of the breech, to facilitate a lateral escape of gas in case of a burst cartridge. There is also a third hole or gas escape in the bolt-head in case of a blow back from a defective cap.

61. The magazine is made from sheet steel, and holds 10 cartridges. The cartridges rest on a platform, which is actuated by a spring made from ribbon steel.

62. To afford a smooth bearing for the platform an auxiliary spring is fixed by a lip and stud to the front of the magazine. On the right there is a stop clip to retain the platform in its place and keep the top cartridge in position. The magazine can be released by pressing the catch which projects through the trigger guard in front of the trigger.

63. A cut-off is provided, which consists of a flat piece of steel hinged to the right side of the body and moved by means of a thumb-piece, which requires a downward and outward pressure of the thumb.

64. The upper band and nose-cap are formed in one piece, to which the piling swivel is attached. One sling swivel is attached to the lower band, the other is screwed into the butt.

65. The butt and fore-end are separate. The butt has a pistol grip and a butt trap cover with spring to enable an oil bottle and pull-through to be carried in the stock-bolt hole. The butt-plate is of yellow metal. The rear end of the body is formed with a socket into which the butt fits; this latter is held in position by the stock-bolt, which screws into the socket of the body. The regiment and number of the rifle are marked on the tang of the butt-plate.

66. The barrel is screwed into the body and secured to the fore-end by the nose-cap and lower band.

67. A wooden handguard is secured over the barrel in rear of the backsight by two spring clips.

68. The foresight is of the blade pattern and is made in five heights (marked ·933, ·948, ·963, ·978, and ·993 respectively) to allow any deviation from the normal in the shooting of individual rifles to be corrected. It is adjustable laterally.

69. The backsight is graduated up to 1,800 yards, the graduation marks being figures and lines, denoting hundreds and fifties of yards respectively. The slide is held in position by a clamping screw with milled head on the right of the slide. In the slide a windgauge is provided, which is adjusted by means of a milled head on the left of the slide. It has a scale, each division on which represents 6 inches per 100 yards.

Inside the windgauge is a friction spring which prevents accidental movement and gives a click for each quarter revolution of the milled head, and 6 clicks, or one and a half revolutions of the milled head, are equivalent to one division on the scale.

70. On the left side of the rifle are provided long range sights for ranges from 1,600 to 2,800 yards. These consist of an aperture sight pivoted to the rear of the body, and a dial sight with pointer secured by means of a screw to the fore-end in rear of the lower band. The ranges are marked on the dial plate.

71. Great care must be exercised when taking off the handguard for cleaning purposes. It will be removed as follows:—Hold the small of the butt between the thighs, place the tips of the forefingers, pointing upwards, under the front end of the hand-guard where rounded off, place the tips of the thumbs on the hinge of the backsight, and press upwards with the forefingers.

72. The action of the mechanism is similar to that of the Short Magazine Lee-Enfield Rifle (Section 3), except that on pressing the trigger when the rifle is at full cock the nose of the sear is released from the full bent of the cocking-piece, and the mainspring carries the striker forward, exploding the charge (para. 40).

73. *To strip the bolt.*—Remove the striker keeper-screw and bolt-head. Using the tool for removing the striker, unscrew the striker, and remove striker, mainspring, and cocking-piece.

Place the safety catch in the midway position, and drive out the finger piece from the left. Remove the plunger and spring.

74. *To assemble the bolt.*—Reverse the above operations, being careful to see that the bolt-head is screwed fully home before replacing the bolt in the rifle.

Section 6.

. (*Cancelled.*)

Section 7.—*The Webley pistol.*

77. i. The principal parts are the barrel, the cylinder and the body. The *barrel* is pivoted to the body by means of a screw passing through a knuckle-joint, formed in a bracket, which projects beyond the breech-end. A strap or rib extends backwards from the top of the barrel; when the barrel is in firing position the rear end of the strap fits on to the body, and is locked down by the barrel catch. The *cylinder* is chambered to hold six cartridges. It revolves on a tube fixed in the bracket under the barrel. The stem of the extractor lies in this tube, surrounded by a spiral spring, which returns the extractor to position after it has ejected the cartridge cases. The extractor is forced out by a small lever in the knuckle-joint, as the barrel, on being pressed downwards, causes the cylinder to rotate on the joint pin.

ii. Length of barrel	4 inches.
" " cylinder	1½ "
" " pistol	10¼ "
Calibre	·441 "
Number of grooves	7.
Rate of spiral	1 in 20 calibres.
Weight of bullet	265 grains.
Charge of cordite	6½ grains.

Size 1/·05

Muzzle velocity	640 f.s.
Weight of packet of 12 cartridges ...	9½ ounces.
Weight of pistol	2 lbs. 3 ozs.

Section 8.—Action of the mechanism, &c.

78. The pistol may be fired in two ways :—

i. By the trigger action, for "continuous practice."

ii. By the cocking action, for "single practice."

79. *Trigger action.*—When using the trigger action, the trigger is pressed back almost to the guard, and released when the shot is fired. The rotation of the cylinder, compression of mainspring, and ignition of charge are carried out automatically as follows :—

i. Assume that one shot has been fired.

ii. On pressing the trigger, the trigger-catch is withdrawn from the small slot on the cylinder ; the pawl, which is connected with the trigger, is made to move upwards and slightly forwards ; its point engages the ratchet on the cylinder (causing the latter to revolve on its axis until a new cartridge is brought into position), and then assists to keep the cylinder in position.

iii. At the same time the cylinder stop, rising through a recess in the body, engages with the wide slot on the cylinder, preventing the latter from rotating too far.

iv. The action of pressing the trigger causes the trigger-nose to rise and engage with the hammer-catch.

v. As the trigger-nose continues to rise, the hammer is rotated on its axis, the top portion swinging backwards, and thereby compressing the mainspring.

vi. When the trigger has been pressed back sufficiently far the trigger-nose disengages from the hammer-catch, and the hammer is then free to fly forward, actuated by the compressed mainspring.

vii. At the same time the trigger-catch, actuated by a spring, engages in the narrow slot in the cylinder. This keeps the cylinder steady, and retains it in position at the moment of firing.

viii. The hammer-nose enters the firing hole, striking the cap and exploding the charge.

ix. On releasing the trigger, it is forced back to its original position by the pressure of the mainspring auxiliary, actuated by the short arm of the mainspring, the cylinder being held by the trigger-catch.

x. The hammer is at the same time brought back to rebound position by the shorter side of the auxiliary bearing in the rebound-arm.

80. *Cocking action.*—When using the cocking action the hammer is drawn back with the thumb till the nose of the trigger engages in the bent on the hammer, and the hammer is held in the firing position. This causes the mainspring to be compressed, the cylinder to be rotated, and the front cylinder to rise. On pressing the trigger the nose is released from the bent, and the hammer actuated by the mainspring is permitted to fly forward and explode the charge as before. The remaining components work as in firing by the trigger action.

81. In both methods, after each shot, the trigger must be allowed to come back freely by releasing the fore-finger as much as possible, without taking it out of the trigger guard. As much greater accuracy can be obtained when

firing single shots slowly than when firing all the chambers rapidly and continuously, the latter method should only be adopted in an emergency.

82. The pistol is sighted to shoot accurately at 50 yards. It has been found by experiment that the extreme range of the bullet fired from the Webley pistol is about 1,550 yards, this range being obtained at an angle of 35°. It should, however, as a general rule, be reserved for use at close quarters.

83. The pistol must be opened and closed gently; and only when the hammer is down. Neglect of this rule may result in—

i. Burring the extractor lever and the body, which will cause faulty extraction and difficulty in loading.

ii. Premature explosion.

iii. Injury to the extremity of the pawl, from which may result partial failure of rotation of the cylinder; and the pawl may not give its assistance in holding the cylinder in the firing position.

Every part of the pistol must be kept perfectly clean, and for this purpose Russian petroleum, and flannelette or old sheeting *only* will be used. The cleaning rod should be passed down the barrel from the muzzle and through each of the chambers of the cylinder, which should not be taken off the cylinder axis.

Care of Arms.*

Section 9.—General instructions.

84. Officers commanding companies are responsible for the condition of the arms on their charge, and for instructing their men in the use of the gauze so that no unnecessary wear of the bore may result.

---

* For the names of the parts of the rifle *see* Plates I and VI.

85. Wear in the bore of a rifle is due to three causes: (*a*) the friction of the bullet; (*b*) the heat generated when ammunition is fired; and (*c*) the friction of the pull-through gauze when the bore is being cleaned.

When care is used in cleaning, 5,000 to 6,000 rounds can be fired from a rifle before it becomes unserviceable.

Undue wear is caused by improper and unnecessary use of the pull-through gauze, to prevent which it is most important that the "Instructions for Cleaning" (Sec. 10) be adhered to. It is recognized that it may be necessary to modify these instructions to suit local climatic conditions, or individual rifles which are in a bad state of preservation.

86. When a rifle barrel is new, the interior of the bore carries a high polish, and this is a great safeguard against rust and metallic fouling, but it must be recognized that, as the bore becomes worn, this polish will diminish. Efforts to restore it with wire gauze on the pull-through result in unnecessary wear. At the same time it must be clearly understood that, in a well-cared-for rifle, while the brilliancy of the polish will diminish, the lands of the bore should still be bright and free from all stain of rust or fouling.

87. A pull-through fitted with a weight, and an oil bottle to contain Russian petroleum, are carried in the recess in the butt of the rifle. The pull-through is packed above the oil bottle as follows:—Hold the pull-through (loop end) between the forefinger and thumb of the left hand, so that the end falls about 2 inches below the third finger; roll it loosely three times round the first three fingers. Slip the coil off the fingers, and lap it tightly with the remainder of the cord, leaving sufficient to allow the weight to drop easily into the recess in the butt. Push the cord into the trap leaving the loop end uppermost, drop the weight into the recess, and drop the trap. The pull-through is made with three loops: the first (*i.e.*, nearest the weight) is for the gauze when used; the second for the flannelette; the third is provided merely as a means of withdrawing the pull-through in case of a jamb; neither flannelette nor gauze should be placed in this loop. When signs of wear appear a new cord should be taken into use, to avoid the risk of the pull-through breaking in the rifle. If a breakage does occur the rifle must be at once taken to the armourer. No attempt should be made by the soldier to remove the obstruction.

88. *Use of the pull-through.*—Remove the bolt from the rifle and, in order to ensure the gradual compression of the gauze, if used, and of the flannelette, drop the weight through the bore *from breech to muzzle.* The pull-through should be drawn through in one motion, otherwise the spot where the flannelette is allowed to rest, while a fresh grip of the cord is being taken, will not be properly cleaned. Very great care must be taken not to allow the cord to rub against the muzzle, otherwise a groove, technically known as "cord wear," will be cut, which in course of time will destroy the accuracy of the rifle.

Only regulation flannelette is to be used. When cleaning or drying the bore after washing out with water, a piece of dry flannelette large enough to fit the bore tightly (about 4 inches by 2 inches) should be placed in the second loop of the pull-through.

For oiling the bore, a slightly smaller piece of oily flannelette, which will fit the bore loosely, should be used. Care must be taken not to use too much oil, as it will be squeezed out of the flannelette at the entrance to the bore and will run down into the bolt when the rifle is placed in the rack, and may then cause miss-fires.

The use of two single pull-throughs attached to one another so as to make a double one, is strictly forbidden, because this practice has been found to produce "cord-worn" barrels.

89. Wire gauze in pieces 2½ inches by 1½ inches is supplied, and should be used for the removal of hard fouling or of rust. In attaching it to the pull-through the following method should be adopted:—Turn the shorter sides of the gauze towards the upper, so that the longer sides take the form "S." Open the first loop of the pull-through (*see* para. **87**), and put one side of it in each loop of the "S." Then coil each half of the gauze tightly round that portion of the cord over which it is placed till the two rolls thus formed meet.

WIRE GAUZE FOLDED.

SECTION

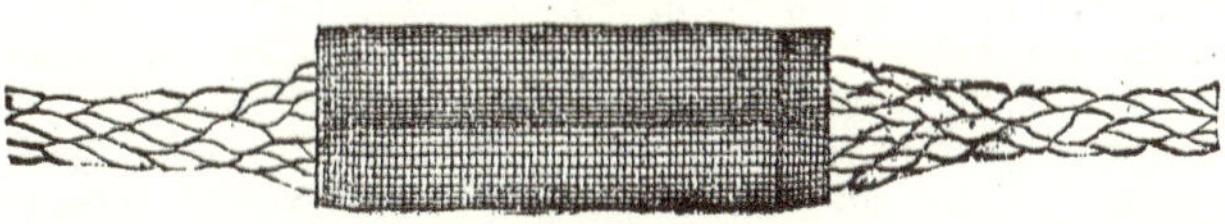

ON PULL-THROUGH.

The gauze must be thoroughly oiled before use to prevent its scratching the bore.

The object of the gauze is mainly to scour out the grooves, and it should therefore fit the bore tightly. When it fails to do this, it should be partially unrolled and packed with paper or flannelette to increase its bulk.

Grit must be removed from the gauze and pull-through before use.

90. Cleaning with gauze entails wear of the bore of the rifle. Gauze should not be pulled through the barrel more often than is here laid down (para. **99**) without sufficient cause. The surest way of preventing the necessity for the continued use of gauze is to keep the bore well oiled so as to prevent rust. A barrel which has become rusty will always be more liable to rust than one which has been kept in good condition. It will therefore require more attention and more frequent cleaning with gauze. Similarly, a barrel in which erosion has commenced will require more care than one of which the surface has not been attacked, for, the eroded portion being rough, moisture is more likely to collect on it and form rust. It is also more difficult to remove rust thoroughly from a rough surface than from a smooth one.

91. No oil other than Russian petroleum should be allowed to remain in the bore. The function of this oil is to cover the bore with a waterproof film, and thus prevent moisture attacking the steel and forming rust. It must be well worked into the flannelette with the fingers, otherwise it will be scraped off by the breech end of the barrel. When paraffin has been used, all traces of it should be thoroughly removed and the bore coated with Russian petroleum, for paraffin, though an efficient agent for removing rust, does not prevent its formation.

92. No gritty or cutting material, such as emery powder or bath brick, is to be used for cleaning any part of the rifle.

Section 10.—*Instructions for cleaning.*

93. In order that the "Instructions for cleaning" may be understood, it is essential that the causes of fouling in rifle barrels should be briefly explained. Fouling may be said to be of two kinds:—(*a*) Internal—probably caused by the forcing of gas or harmful material into the pores of the metal; (*b*) Superficial—caused by the deposit in the bore of the solid products of combustion of the charge and of the cap composition.

The result of neglect in either case is the same, viz., the formation of rust in the bore and, as a consequence, corroded barrels, calling for the excessive use of wire gauze, or even more drastic treatment, thereby causing unnecessary wear.

Internal fouling can be removed satisfactorily by the use of boiling water (*see* para. **101**). It for any reason this method of cleaning cannot be used, the barrel will "sweat," and a hard black crust of fouling will appear in the bore. This will turn to red rust if not removed, and the rifle will then require repeated cleaning with flannelette and probably with gauze, for a time which will vary according to climatic conditions and the state of the bore.

Superficial fouling is readily removed when warm by the use of a pull-through and flannelette, but if it is allowed to remain long in the barrel, it will become hard and will have a corrosive effect equal to that produced by internal fouling.

94. *To remove the bolt.*—Raise the knob as far as it will go, draw back the bolt-head to the resisting shoulder, and release it from the retaining spring. Raise the bolt-head as far as possible (in the short rifle, Marks I, I*, II, and II*, draw back the charger guide, then turn the bolt-head to the left) and remove the bolt.

95. *To replace the bolt.*—See that the resisting lug and cocking-piece are in a straight line, and the bolt-head screwed home. Place the bolt in the body with the extractor upwards, and press it forward until the head is clear of the resisting shoulder (in the short rifle, Marks I, I*, II, and II*, turn the head to the right, then push the charger guide forward as far as possible). Turn the head downwards until it is caught by the retaining spring. Close the breech and press the trigger.

In some rifles, the bolt can be replaced and closed with the bolt-head unscrewed a whole turn. It cannot, however, be closed with the bolt-head in this position if there is a cartridge in the chamber. The greatest care should therefore be taken to see that the bolt-head is screwed fully home before the bolt is placed in the rifle.

96. *Daily cleaning.*—The outside of the rifle will be cleaned daily, all parts of the action wiped with an oily rag; the bore of the rifle will always be left oily—once a week this oil will be removed and the bore re-lubricated. In the case of rifles that have once become rusty, the bore will be daily wiped out with flannelette and re-oiled; it will in addition be cleaned once a week with the gauze on the pull-through. The gauze to be packed as directed in para. **89** so as to fit the bore tightly. For daily cleaning after firing, *see* para. **99.**

97. *Cleaning before firing.*—The action will be wiped with an oily rag, and all traces of oil will be removed from the bore and chamber by the use of a pull-through *which has no gauze on it.*

Neither the cartridge nor the chamber of the rifle are, on any account, to be oiled before loading, nor is any other form of lubricant to be used with a view to facilitate the extraction of the empty case. Such a procedure greatly increases the thrust on the bolthead due to the explosion of the charge and is liable to injure the rifle.

98. *Cleaning after firing.*—Arms will be cleaned immediately after firing. The fouling can be easily removed while it is still warm, and before it has had time to set hard, while the less the time that is allowed for the fouling to exercise its power of absorbing moisture from the air, the less chance is there of rust forming. If it is impossible to clean the rifle at once, an oily rag should be pulled through the bore, and the rifle should be cleaned at the earliest opportunity.

99. *The bore.*—The following method of cleaning the bore should be adopted :—

Thoroughly oil the gauze to prevent its scratching the surface of the metal, drop the weight of the pull-through through the bore from the *breech,* and pull the gauze through three or four times. Then place a tightly-fitting piece of dry flannelette in the second loop of the pull-through (*see* para. **87**), and draw it through till the bore is clean. Finally oil the bore with a loosely fitting piece of flannelette, using enough oil to cover the bore thoroughly. The rifle will be cleaned in this manner for three days following that on which it was fired.

100. After firing blank ammunition, special care should be taken that the cleaning is thorough, as, although there is no friction between bullet and bore and so no internal fouling or "sweating," there is greater accumulation of superficial fouling from blank than ball cartridge, there being no bullet in blank ammunition to scour the fouling left by the preceding round. The firing also is in most cases more prolonged, and a greater interval must usually elapse before the rifle can be thoroughly cleaned. When blank firing precedes practice with ball, the rifles will be carefully cleaned before ball practice commences.

101. An effective means of cleaning the bore, whether firing has taken place or not, is found in the use of boiling water. Before boiling water is used superficial fouling and grease should be removed. About five or six pints should be poured through the bore from the breech, using a *funnel to prevent its entering the body or magazine.* The rifle should then be thoroughly dried and the bore oiled. Not only does the boiling water remove the fouling, but the expansion of the metal due to the heat of the water loosens any rust there may be and makes it easily removable.

102. The appearance of nickelling or metallic fouling should be watched for. It is caused by a portion of the cupro-nickel of the envelope of the bullet being left on the surface of the bore, and appears as a whitish streak on the lands, or as a slight roughness on the edge of the grooves. If it is deposited near the muzzle or the breech it is visible to the eye when the bore is clean, but in the centre of the bore it can only be detected by the use of the gauge plug. It is a cause of inaccuracy and if a rifle for no apparent reason shoots badly, its presence should be looked for as a possible explanation. The soldier will make no attempt to remove it himself, but will hand his rifle to the armourer, or other qualified person, to be cleaned.

103. *The action and outside.*—Thoroughly clean the bolt, paying particular attention to the face of the bolt-head, the striker point, and the extractor. If the bolt requires cleaning inside it will be taken to the armourer.

See that the recess for the extractor spring is clear of dirt. Take out the magazine and wipe the inside of the body and the entrance to the chamber with an oily rag. Remove all dirt from the slots in the charger guide and from the extractor recess in the front of the body. Take out the magazine platform (para. **47**) if required and clean the inside of the magazine with a dry rag. Wipe the exterior of the rifle with an oily rag, seeing that the "U" of the backsight, the hole in the aperture sight, the gas escape holes and, in the short rifle, the rack on the side of the leaf, are free from dirt. Remove any fouling which has collected on the bayonet boss on the nose-cap. If allowed to accumulate, this may cause difficulty in fixing the bayonet.

104. The instructions regarding the use of an oily rag for cleaning the bolts and bodies will not apply in dusty countries, where all parts of the action will be kept dry and clean.

105. *Cleaning ·22-inch rifles and aiming tubes.*—As a foul rifle shoots very inaccurately it is of the utmost importance, from considerations of safety, that the barrel should be frequently wiped out during use.

The rod and brush should be inserted from the breech end. Under no circumstances should they be inserted from the muzzle, as the friction of the rod is liable to enlarge the bore and make the muzzle bell-mouthed, thus causing inaccuracy.

Section 11.—Instructions for care of arms and ammunition.

106. When the rifle is not in use, the leaf and slide of the backsight should be lowered to avoid the risk of damage from a blow or fall.

107. The mainspring should never be allowed to remain compressed, except when the rifle is loaded, as the spring will thereby be weakened. The position of the cocking-piece shows whether the mainspring is compressed or not.

108. The pull-off is the amount of pressure which is required to release the nose of the sear from the full bent of the cocking-piece; it should not be heavier than 6 nor lighter than 5 lbs. in the short rifle; in other rifles, not heavier than 7 nor lighter than 5 lbs. Any defect therein should be remedied by the armourer only.

109. The magazine must not be removed from the rifle except for cleaning purposes, and, to avoid weakening the spring, cartridges should only be kept in it when necessary. A failure of the spring to raise the platform can usually be overcome by tapping the bottom of the magazine smartly with the palm of the hand. If the failure recurs, the rifle should be taken to the armourer for examination and repair.

110. The bolts of rifles are not to be exchanged. Each bolt is carefully fitted to its own rifle, so that the parts which take the shock of the explosion have an even bearing, and the use of a wrong bolt will affect the accuracy of the rifle. The number stamped on the back of the bolt lever should agree with that stamped on the right front of the body.

111. Care should be taken to prevent the browning being rubbed off the rifle.

112. No non-commissioned officer or soldier is permitted to take to pieces any portion of the action, except as prescribed for cleaning, nor is he allowed to loosen or tighten any of the screws, unless authorized to do so by his company commander.

113. In dusty countries it may be found necessary to cover the muzzle and bolt with a cover of khaki or other suitable material, to prevent the dust gaining access to the interior of the rifle, but anything in the nature of a plug in the muzzle is expressly prohibited.

114. A miss-fire arises from :—

i. A defective cartridge.
ii. A defective rifle.

In Case i. the cartridge will be tried in another rifle, and, if it still fails to fire, a report will be made in accordance with the instructions contained in the "King's Regulations."

In Case ii. the rifle will be taken to the armourer for examination.

115. The oil will only be removed from the bore of the rifle :—

i. Immediately before firing.
ii. For inspection, which, except after firing, should not, as a rule, be more often than once a week (*see* para. 96).
iii. For parades and duties as may be ordered by the commanding officer.

In all cases it will be replaced as soon as possible.

After firing with bayonets fixed, the bayonet should be carefully wiped before it is returned to the scabbard. All oil should be removed from the blade before placing a bayonet in the scabbard.

116. Ammunition should be kept perfectly dry and clean, and should not be exposed to extremes of temperature.

117. Commanding officers will report in the Regimental Annual return (para. **620**) any defects in the machine guns, rifles, or ammunition on their charge, which have not been satisfactorily remedied.

Section 12.—*Examination of small arms.*

118. It is necessary for all company officers and serjeants to possess a competent technical knowledge of the inspection, care, and preservation of small arms. Commanding officers will therefore arrange that they shall be instructed annually by the regimental armourer in repairing faults most likely to occur in the field with such tools as would be available, and in the examination of the various components as directed in the following paras. :—

Rifles, M.L.M.

i. The interior of the barrel for rusts and cuts.

ii. (*a*) The backsight leaf, for firmness of joint; that it is not bent; that the slide moves smoothly and fits firmly on the leaf; that the V is not deformed, and that the lines on the slide are clearly marked.

(*b*) The foresight, that the barleycorn is not deformed.

iii. The aperture and dial sights; that they are not bent, and work smoothly.

iv. The bolt cover; for security on the bolt and clearance of the body.

v. The cocking piece; for firmness on the striker, that the bents are in good condition, and that the sear nose bears properly.

vi. The sear; for height of the nose, which should just clear the bottom of the resisting lug on the bolt.

vii. The butt; that the stock bolt is properly screwed up.

Note.—In arms marked 2 on the right of the butt, and at the socket of the fore-end, the latter must be removed before attempting to turn the stock-bolt. In screwing it home the precautions prescribed in sub-para. xvii. will be carefully observed.

viii. The cocking-piece and striker; that they fly forward freely on pressing the trigger.

ix. The striker point; that it is the correct shape and projects sufficiently through the face of the bolt-head.

x. The magazine; that it is not dented, and that the platform works freely.

Rifles, M.L.E.

The same as for rifles, M.L.M., with the following addition:—

xi. Safety catch; that the bolt of the safety catch engages in the slots in the extension at the end of the bolt.

Rifles, M.L.E. Charger-loading.

The same as i., iii., v., vi., vii., viii., ix., x. and xi., with the following addition:—

xii. (*a*) The backsight leaf, for firmness of the joint; that it is not bent; that the slide moves freely; that the clamping screw engages properly; that the windgauge fits firmly; and that the U is not deformed.

(*b*) The foresight; that the blade is not deformed.

Rifles, Short, M.L.E., Marks I and I and Converted Marks II and II*.*

The same as i., iii., v., vi., viii., ix. and x., Rifles, M.L.M., with the following additions :—

xiii. (*a*) The backsight leaf, for firmness of the joint; that it is not bent; that the fine adjustment and windgauge fit firmly; that the slides move smoothly; that the catches engage in the racks on both sides of the leaf; and that the V is not deformed.

(*b*) The foresight; that the barleycorn is not deformed.

xiv. The bolt; that the striker is not too free on the cocking-piece and that it is not screwed too far into the latter; also that the striker keeper nut screw (or the striker keeper-screw if fitted) is not broken, and that the nut is in its proper position.

xv. The bolt head; that the charger guide is not too loose on the bolt head; that it works smoothly; and that the top screw is intact.

xvi. The safety catch and locking bolt; that the safety catch engages in the camway of the bolt and locks it; that it does not move too easily; and that the cocking-piece is withdrawn slightly to the rear when the locking bolt is applied, whether it is at "full cock" or the "fired" position.

xvii. The butt; that it is not loose. If the stock bolt requires screwing up to tighten the butt, the fore-end must first be removed. On reassembling, great care will be taken that the square end of the stock bolt which protrudes through the socket of the body is in the correct vertical position, so that it may enter the keeper plate properly when the fore-end is replaced. On replacing the fore-end see that the fore-end stud and spring, where fitted, are in proper position; the front guard and inner band screws must be carefully tightened.

Rifles, Short, M.L.E., Mark III, and Converted Mark IV.

The same as i., iii., vi., viii., ix., x., xiv., xvi., and xvii., with the following additions:—

xviii. (*a*) The backsight leaf, for firmness of the joint; that it is not bent; that the windgauge fits firmly; that the slide moves smoothly, that the thumb piece and fine adjustment worm work freely and engage in the rack on the side of the leaf; and that the U is not deformed.

(*b*) The foresight; that the blade is not deformed.

These instructions apply also to Drill Purpose rifles. As these are used for the instruction of recruits in aiming, particular attention should be paid to the state of the sights.

119. Non-commissioned officers and men are to be fitted with rifles having long, normal, or short butts, according to their build and preference. The choice should be made after tests carried out in the standing and prone positions, and should be based on the readiness with which the firer brings his rifle up to the firing position and aligns his sights without letting his nose and mouth come into close proximity to the thumb and fingers of the right hand. The principal consideration is the distance of the nose from the shoulder, and it is, as a rule, the broad-shouldered and long-necked men that require the long butts. A bad trigger release, due to incorrect holding of the small of the butt, will often result from the use of too short a butt.

The butt selected should be the shortest which can be used comfortably when firing both standing and lying down in the correct positions.

Section 13.—*Testing arms.*

120. Officers commanding companies will always report any serious defect which may be discovered in any rifle on their charge to the officer commanding their battalion.

121. A rifle is said to be inaccurate when a number of shots fired from it without alteration of aim or sighting by a skilled man using reliable ammunition and in still atmosphere, strike a large target at several points widely distant from each other.

122. A rifle may also be over-sighted or under-sighted so that shots strike the target consistently above or below the point aimed at, when a regulation aim is taken; or it may have a "throw" right or left, causing the bullets to be deflected to the right or left of the point of aim.

123. A soldier should not be expected to make considerable allowances in sighting for the error of his rifle, and steps should always be taken to remedy such defects.

124. When there is reason to believe that a rifle is inaccurate it will be tested regimentally in the following manner, and diagrams showing the results of the trial will accompany the commanding officer's report to the Chief Ordnance Officer.

125. In conducting tests the following rules must be carefully observed in order to ensure that reliable data are obtained as to the condition of the rifle under trial:—

i. A favourable day will be chosen.

ii. The rifle will be carefully cleaned and examined by the armourer, who will note any defect in the barrel or fore-end, and see that the barrel is not unduly gripped by the nose-cap or bands.

iii. It will be fired with its own bolt.

iv. A skilled shot will be selected.

v. A reliable batch of ammunition will be used.

vi. The firing will be from the shoulder in the prone position, with the wrist supported on sandbags or other suitable rest.

vii. A square white target of 6 feet side, with any suitable black aiming mark, will be used, and the distance will be 500 yards.

viii. Trial rounds will be fired from the rifle to determine the elevation.

ix. Aim will be taken as directed in the rules for aiming, and the slide of the backsight will be adjusted so as to ensure that all shots strike the target. The fine adjustment will not be employed. With the short M.L.E. rifle, Mark III or IV, the slide will be adjusted at one of the graduations on the leaf of the backsight. With L.M. and L.E. rifles the slides will be placed at one of the graduations, not between two lines. The elevation used will be noted on the diagrams.

x. When the elevation has been ascertained, the rifle will be cleaned and three shots to warm the barrel will be fired into the butts, followed by 7 rounds with the same aim and elevation as in ix. The force of the wind will be ascertained by means of a rifle, known to shoot well, fired by a skilled shot; and the amount of deflection due to the wind will be entered on the diagram.

xi. On the completion of the firing the point aimed at and the hits on the target will be transferred with the greatest care to Army Form B 202, which is termed the diagram (para. 126). Hits will be numbered from 1 to 7, according to the order in which they were fired. Should a miss occur, the diagram will be commenced again.

xii. Three diagrams will be fired with each rifle to be tested.

xiii. The ammunition must not be exposed to extreme heat or cold, nor should a cartridge be inserted in the chamber until all is ready for firing a shot.

126. The "Figure of Merit" of the rifle will then be computed in the following manner:—

The distance of the centre of each hit will be measured from the left edge and from the bottom of the target. The measurements will be in feet, to two places of decimals, and will be at once entered in the proper columns on Army Form B 202. Similar measurements in respect of the mark aimed at will be entered in the space provided. Each column will then be totalled and the means found. The intersection of the lines representing the mean horizontal and the mean vertical positions of the group will give the position of the point of mean impact. Each shot will then be plotted on the diagram from its measurements and the point of mean impact (marked ⨀), and the mark aimed at (marked ×) will also be shown. The deviation of each shot from the point of mean impact will then be measured and entered in the proper column. The mean of the deviations gives the "Figure of Merit" of the rifle.

The "Figure of Merit" will show the accuracy of the rifle. When the average of the figures of merit of three diagrams exceeds 1 foot the rifle will be reported inaccurate.

127. Full particulars regarding the mark, source of manufacture and date of issue, &c., of the rifle, and the mark, date, and place of manufacture of the ammunition must be given on the diagram.

128. The extent to which a rifle is incorrectly sighted in a lateral sense may be found by measuring from the point of mean impact to a line drawn vertically through the point aimed at, and then adding or deducting the allowance due to wind.

129. A vertical error will be demonstrated partly by the elevation required, and partly by measuring from the point of mean impact to a line drawn horizontally through the point aimed at.

130. Should a rifle, otherwise accurate, be reported as over-sighted or under-sighted to such an extent that its value is seriously impaired for practical purposes, the commanding officer should arrange for its sighting to be tested by a skilled shot, under favourable conditions of weather, at 500 yards. If he is satisfied that the defect, as ascertained, can be remedied by the substitution of a higher or lower foresight than that used in the test, he may direct the armourer to carry out the alterations, but care should be taken that the regulation aim is taken in the test, and that the change of foresights is made by the armourer.

131. A lateral error can also be remedied by the armourer. The graduations on the windgauge afford a guide as to the amount of correction required.

132. If, after re-adjustment and a second test, a rifle is still found to be unsatisfactory, the barrel and body will be exchanged.

132A. Rifles are examined periodically by the Army Inspection Department, the object of such examination being to ensure that the arms are in a condition to withstand the wear and tear of a campaign. The mere fact of a rifle being rejected as "unserviceable" on examination does not therefore imply that it is necessarily "inaccurate."

Ammunition.

Section 14.—*Cartridge, S. A. Ball* ·303 *inch* (*Mark VI*).

133. Charge—About 31 grains of cordite, in 60 small strands.

Bullet—Compound, consisting of a core containing 98 per cent. lead and 2 per cent. antimony, in an envelope of 80 per cent. copper and 20 per cent. nickel.

A shallow cannelure is formed r ear the base, into which the case is secured by three indents. The diameter of the bullet at base is ·311 in. Its length is 1¼ in.

The portion of the bullet which enters the case is coated with beeswax.

Case—Of brass, solid drawn, 70 per cent. copper and 30 per cent. zinc.

Wad—A glazed cardboard disc is placed on top of the charge.

Weight—Of bullet, 215 grains.
,, Of cartridge, 415 grains.
,, Of one packet of 10 rounds, 10 ozs.
,, Of one cardboard box containing 20 rounds in chargers, 22 ozs.
,, Of one box of 1,000 rounds in chargers, 80 lbs. 10 ozs.

Cartridge, S. A. Ball ·303 *inch* (*Mark VII*).—Charge—About 38 grains of tubular M.D. cordite in about 40 small tubes.

Bullet—Compound, consisting of a core in two parts—the front portion being an alloy of aluminium 90 per cent., zinc 10 per cent., or pure aluminium, and the rear portion being an alloy of 98 per cent. lead, 2 per cent. antimony, in an envelope of 80 per cent. copper and 20 per cent. nickel.

A shallow cannelure is formed near the base, into which the case is secured by three indents. The diameter of the bullet at the base is ·311 inch and its length 1·28 inch.

The portion of the bullet which enters the case is coated with beeswax.

Case—Of brass, solid drawn, 70 per cent. copper and 30 per cent. zinc.

Wad—A glazed cardboard disc is placed on top of the charge.

Weight—Of bullet, 174 grains.
,, Of cartridge, 382 grains.
,, Of one packet of 10 rounds, 9 ozs.

Weight—Of one bandolier, cotton, containing 50 rounds, 10 rounds in chargers, in each packet, 3 lbs. 2½ ozs.

" Of one box of 1,000 rounds in bandoliers, 73 to 74 lbs.

Section 15.—*Allowance of ammunition.*

134. The scales of ammunition sanctioned by the Equipment Regulations are not to be exceeded. Nothing in these regulations is to be regarded as sanctioning excess expenditure of ammunition.

135. The allotment for the Regular Forces may be drawn for each recruit and trained soldier who begins to fire Table A or Table B.

136. Ammunition for repetition of the elementary instructional practices, Part I, Table A, of recruits of cavalry and infantry and Royal Engineers (Sappers) of the Regular Army may be demanded in addition up to the full amount required.

137. The allowance of ammunition made to general officers and to officers commanding for purposes of inspection and testing accuracy of scoring is given in Equipment Regulations, Part I, 1909.* General officers commanding-in-chief are, in addition, authorized to reserve 10 rounds per man of the annual allowance of ball ammunition to enable them to test by inspection the execution of range and field practices.

---

* For instructions as to the return of empty cartridge cases and chargers, *see* Equipment Regulations, Part I, 1909.

138. Officers, non-commissioned officers, and men, of the regular establishment, posted for a short tour of duty to Special Reserve battalions, will draw the same amount of small-arm ball ammunition as is allowed for Regular Troops of the branch of the service to which they belong, except the ammunition prescribed for collective field practices. They will fire the Trained Soldiers' Course prescribed for their regular unit, with the exception of the collective field practices, the allowance of ammunition for field practices being reduced in their case to 35 rounds per man. Those who are posted permanently to Special Reserve battalions will draw the same allowance of small-arm ball ammunition as is laid down for the trained Special Reservists, and will fire the annual course prescribed for the Special Reserve.

139. The rounds shown as "surplus" in Tables A and B, will be distributed primarily by commanding officers for the following purposes:—

i. Repetition firing as prescribed in the tables.
ii. 15 rounds per man to be expended by company commanders for further training of indifferent shots.
iii. The further training of recruits (para. **407**).
iv. Fire direction practices.
v. The testing of rifles when necessary (paras. **124** and **379**).
vi. Occasional shots on the rifle range (para. **460**).
vii. 10 rounds per man who commences the course for voluntary practice.

Any surplus available after the above requirements have been fulfilled will be distributed to companies for general practice.

It is not to be used for practice for competitions.

140. The allocation of this ammunition will be based on the principle that the object to be attained is a high general standard of shooting, for which purpose there should be frequent opportunities for practice throughout the year. The use of ammunition purchased privately immediately prior to or during classification practices is opposed to the spirit of these regulations, and should not be permitted, except in the case of the Special Reserve and Territorial Force.

141. A daily record (Army Book 99, 100 or 218) will be kept by each company commander and assistant adjutant or instructor of musketry, in the Regular Forces, Special Reserve, and Territorial Force, of all ammunition expended under the various headings, and the ammunition account in the company annual return will be compiled from it (para. **632**).

A record of ammunition expended in machine gun practice will be kept by the machine-gun officer in the "Machine Gun Register and Diary," Army Book 107 (para. **636**).

These records will be called for and examined by accountants from time to time and will also be inspected by commanding officers when striking off ammunition as expended.

142. Any ammunition remaining on hand at the end of the year will be carried forward in the ammunition account, and deducted from the amount to be drawn in the following year.

143. No cartridges, whether ball, blank, miniature, or dummy, other than those supplied by Government, may be used in service rifles.

Section 16.—*Defective ammunition.*

144. When there is reason to consider ammunition defective the procedure laid down in King's Regulations will be adopted.

145. In making reports of failures which occur with small-arm ammunition, the following definitions will be used :—

i. Burst cases.
ii. Separations.
iii. Split cases.
iv. Fluted cases.
v. Blow backs.

i. Burst cases, as distinguished from separations, may be of two kinds, viz., circumferential or longitudinal, and in reporting them, their position, whether in or above the base, should be clearly stated. A burst well away from the base is less likely to be serious than one in, or close to, the base.

ii. Separations are failures which are due to the case being stretched on firing, owing to excessive backward play of the lock or bolt-head in the Maxim gun or rifle respectively. Separations may be partial or complete, and may take place at any part of the case. They are distinguished from bursts by the torn edges of the metal not being fused. In case of doubt as to whether the casualty is a " burst " or a " separation," the rifle or machine gun should be overhauled.

iii. Split cases are those which burst at the neck or shoulder.

iv. Fluted cases are those in which the powder gas has penetrated between the neck of the case and the walls of the chamber and has forced the metal inwards.

v. A blowback is an escape of gas between the cap and the sides of the cap chamber. This term is not to be used to denote an escape of gas due to bursts or other causes.

CHAPTER III.

THE THEORY OF RIFLE FIRE AND ITS PRACTICAL APPLICATION.

Section 17.—*Application of theory.*

146. A knowledge of the theory of rifle fire is of great importance in enabling the best use to be made of the powerful and accurate weapon with which the soldier is armed, but it is of equal importance that such knowledge should be correctly applied. The moral conditions under which fire is delivered in war are very different from those of peace. It is therefore essential that deductions, made from the theory of musketry or from the results of fire delivered under the conditions of peace, should be considered in the light of careful study of the circumstances of the modern battlefield.

Section 18.—*Definitions.*

147. The *axis of the barrel* is an imaginary line following the centre of the bore from breech to muzzle.

The *line of departure* is the direction of the bullet on leaving the muzzle, *i.e.*, the prolongation of the axis of the barrel.

The *line of fire* is a line joining the muzzle of the rifle and the target.

The *line of sight* is a straight line passing through the sights and the point aimed at.

The *culminating point* is the greatest height above the line of sight to which the bullet rises in its flight; this is reached at a point a little beyond half the distance to which the bullet travels.

The *first catch* is that point where the bullet has descended sufficiently to strike the head of a man whether mounted, standing, kneeling, lying, &c.

The *first graze* is the point where the bullet, if not interfered with, will first strike the ground.

The *dangerous space* for any particular range is the distance between the *first catch* and the *first graze.*

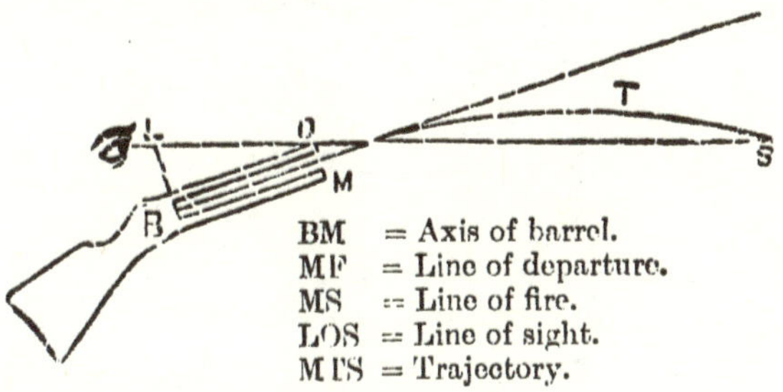

147a. The following definitions are required for machine-gun work only :—

The *angle of tangent elevation* is the angle between the axis of the bore and the line of sight.

The *angle of sight* is the angle between the line of sight and the horizontal plane. It may be either positive or negative according as the target is above or below the firing point.

The *angle of quadrant elevation* is the angle between the axis of the bore and the horizontal plane.

The *angle of departure* is the angle between the line of departure and the horizontal plane.

The *angle of jump* is the angle between the line of departure and the axis of the barrel before firing.

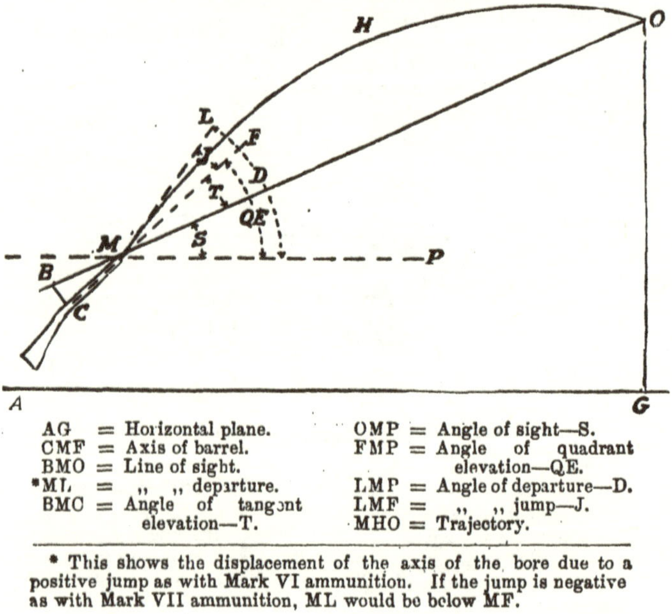

AG	=	Horizontal plane.	OMP	=	Angle of sight—S.
CMF	=	Axis of barrel.	FMP	=	Angle of quadrant elevation—QE.
BMO	=	Line of sight.			
*ML	=	,, ,, departure.	LMP	=	Angle of departure—D.
BMC	=	Angle of tangent elevation—T.	LMF	=	,, ,, jump—J.
			MHO	=	Trajectory.

* This shows the displacement of the axis of the bore due to a positive jump as with Mark VI ammunition. If the jump is negative as with Mark VII ammunition, ML would be below MF.

Section 19.—*Rifling.*

148. A gun barrel is said to be rifled when it has spiral grooves cut down the "bore." Rifling a barrel enables an elongated bullet to be used; the advantage of this form of bullet is that it has great weight in proportion to the surface directly opposed to the air; it has therefore great power of overcoming the resistance of the air, and thus keeping up its velocity. When the charge is fired, the bullet is forced into and follows the grooves up the barrel, thus leaving the muzzle with rotation on its longer axis. This tends to keep its point foremost and therefore to ensure accuracy of flight.

Section 20.—*Forces acting on the bullet.*

149. Three forces act on the bullet: the explosion of the charge, gravity, and the resistance of the air. The explosion of the charge drives the bullet forward. Gravity, *i.e.*, the natural attraction which draws all unsupported bodies towards the centre of the earth with ever-increasing velocity, acts on the bullet immediately it leaves the muzzle. The resistance of the air causes the velocity of the bullet to decrease rapidly.

150. The combined effect of these forces causes the bullet to travel in a curved line called the *trajectory*, the curvature of which becomes more pronounced the longer the bullet is exposed to their action. Thus a Mark VI bullet leaving the muzzle of a service rifle at the rate of about 2,060 feet per second, falls about 4½ inches below the line of departure in the first hundred yards; this drop is increased to about 20 inches at 200 yards. With Mark VII ammunition, giving a muzzle velocity of 2,440 feet per second, the drops are about 3 inches and 13 inches respectively.

151. In explaining the trajectory to recruits, it is not sufficient to draw a diagram representing a trajectory distorted in respect of height and range, nor to throw any object to a distance as a practical illustration. It is most desirable, in addition, to show the path of the bullet at various ranges, say 400 and 800 yards, by means of discs raised on poles at every 100 yards, or some similar device.

Section 21.—*Elevation.*

152. In order to allow for the fall of the bullet it is necessary to direct the line of departure as much above the object to be hit as the bullet will fall below it if the axis of the barrel of the rifle is pointed at the mark. This raising of the barrel to allow for the curve of the trajectory is termed *giving elevation.*

The target must of necessity be kept in view; the rifle is therefore provided with sights which permit the firer to give the elevation required whilst keeping his eye fixed on the mark.

Section 22.—*Sighting of rifles.*

153. In the sighting of rifles a "mean" graduation for each range has been adopted, and a high general standard of accuracy for all practical purposes is thus obtained. Each rifle is carefully tested before issue, but it must be understood that no two rifles behave in an exactly similar manner, and that even if compensation could be made for every error in the sighting of the rifle before issue, wear of parts and loosening or tightening of screws, &c., would bring about faults from time to time. It is therefore necessary that every man should study the shooting of his own rifle, and make himself acquainted with any incorrectness of the graduations marked on the backsight, in order that he may be in a position to give his rifle the correct elevation for the estimated or ascertained range of the target.

154. At longer ranges the backsight elevation may be regarded as the best possible guide under all conditions, or any error may be ascertained by using a long range sighting target.

Section 23.—*Jump.*

155. Owing to the shock of discharge a vibratory or wavy motion is set up in the barrel, and at the moment the bullet leaves the bore the muzzle is usually deflected from its original axis. It therefore rarely happens that the line of departure coincides with the axis of the barrel before firing and the angle between the two is known as the angle of jump. Jump may be either positive or negative according as the muzzle is deflected upwards or downwards with reference to the axis of the barrel. The more serious changes in jump are produced by changes in muzzle velocity, *e.g.*, in the short rifle jump is positive with Mark VI ammunition, whose muzzle velocity is 2,060 feet per second while it is negative with Mark VII ammunition which gives a muzzle velocity of 2,440 feet per second.

Less serious variations are caused by small differences in the stocking of individual rifles and to overcome them, without interference with the standard form of backsight, varying heights of foresight are provided. In the short rifle, each ·015 inch difference in height of foresight makes a difference of about 2½ minutes in the angle of departure.

Lateral jump has also to be considered, but as variations can be readily allowed for by adjusting the position of the fore-sight to the right or left, it is of less importance.

Section 24.—Drift.

156. Drift is the term used to express the lateral deviation of the bullet after it has left the barrel. This deviation, which is considerably less than that caused by jump, is brought about by the rotation of the bullet and the position which it assumes in its flight. The left-handed rifling of the service rifle causes the bullet to rotate from right over to the left, and, owing to gyroscopic action, the point works over slightly to the left. The consequent increased air-pressure on the right side of the bullet therefore forces it to the left.

The deflection due to drift at distances below 1,000 yards is negligible. At 1,500 yards it may be regarded as about 7 feet.

157. The causes and extent of jump and drift are more fully dealt with in the Text Book of Small Arms.

Section 25.—Effect of fixing the bayonet, of resting the rifle, and of a heated or oily barrel.

158. When the bayonet is fixed to the muzzle of the rifle, its weight checks the movements and vibration, called the "jump," and in consequence affects the position of the muzzle at the moment of the departure of the shot, and the primary direction given to the bullet.

159. With Mark VI ammunition the accuracy of the short magazine Lee-Enfield Rifle is not appreciably affected by fixing the bayonet; with Mark VII ammunition an alteration in elevation is required as shown in the following table:—

I. Sight placed at correct elevation to hit the mark without bayonet at	II. Rise of point of impact of bullet after fixing bayonet and using sighting as in Column I.	III. Decrease in elevation required to hit the mark after fixing bayonet.
600 yards	About 4 feet	150 yards.
500 "	" $3\frac{1}{2}$ "	175 "
400 "	" 3 "	200 "
300 "	" 2 "	No decrease possible. It is necessary to aim down.
200 "	" $1\frac{1}{2}$ "	

160. The following table illustrates the effect of the fixed bayonet when firing Mark VI ammunition from the Lee-Metford or Charger-loading Lee-Enfield rifles. With Mark VII ammunition, fixing bayonets has a negligible effect:—

I. Sight placed at correct elevation to hit the mark at	II. Drop of bullet after fixing bayonet and using sight as in Column 1.	III. Extra elevation required to hit the mark after fixing bayonet.
600 yards ...	About 6′ and 2′ to the right	About 100 yards.
500 " ...	" 5′ " 1′ " "	" 150 "
400 " ...	" 4′ " 1′ " "	" 150 "
300 " ...	" 3′ " 1′ " "	" 150 "
200 " ...	" 2′ " 6″ " "	" 150 "

161. It will be seen from Plate XI and from the Trajectory Table (page 67) that, with the backsight adjusted for the distance in question, allowance having been made for fixing bayonets, if necessary, a Mark VI bullet does not rise above the height of a man on foot at 500 yards range, or above the height of a mounted man at 600 yards range. Similarly with Mark VII ammunition the ranges are 600 and 700 yards respectively. It is therefore evident that against a rush of savages or a charge of cavalry, effective fire can be maintained within these ranges without alteration of the backsight provided that the aim is taken at the ground line.

162. For practical purposes the shooting of the rifle is not affected by resting the muzzle, or any portion of the stock, lightly, on earth or other substance.

163. The first round fired from an oily barrel is liable to follow an erratic course, the rifle throwing sometimes high, sometimes low, and at other times to the right or left; a dry rag should therefore be passed through the bore before practice is commenced.

Section 26.—*Dangerous space.*

164. Plate XI illustrates the manner in which the dangerous space decreases as the range increases, the reduction being due to the steeper angle at which the bullet descends at the longer ranges.

165. The extent of the dangerous space further depends on :—

i. The firer's position and the consequent height of his rifle above the ground.

Plate XI.

Comparative Trajectories of Marks VI and Mark VI shewn in Black,

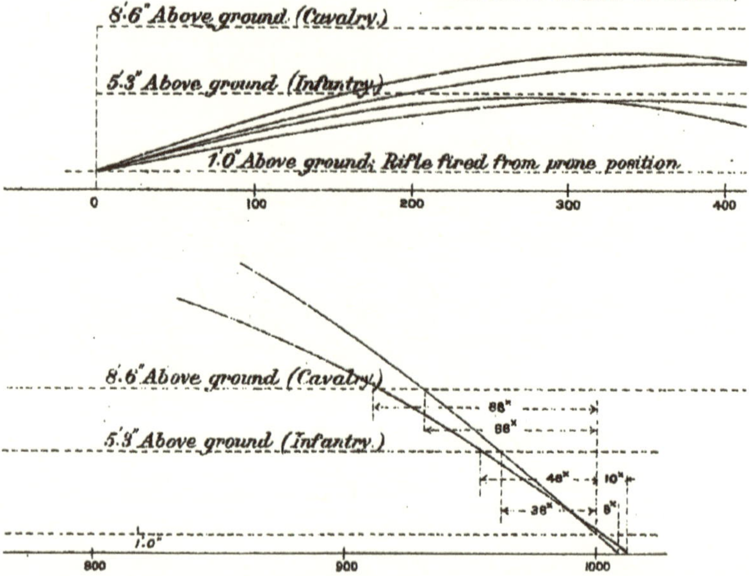

Rifle fired from prone position 1'0" above ground.

4273. 20878/132. 200,000. 8./15.

To face p. 64.

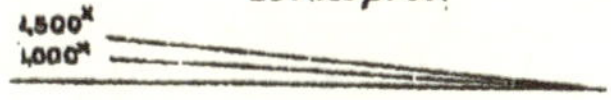

ACTUAL ANGLES OF DESCENT.
(Mark VII ammunition)

VII AMMUNITION AT VARIOUS RANGES.
MARK VII IN RED.

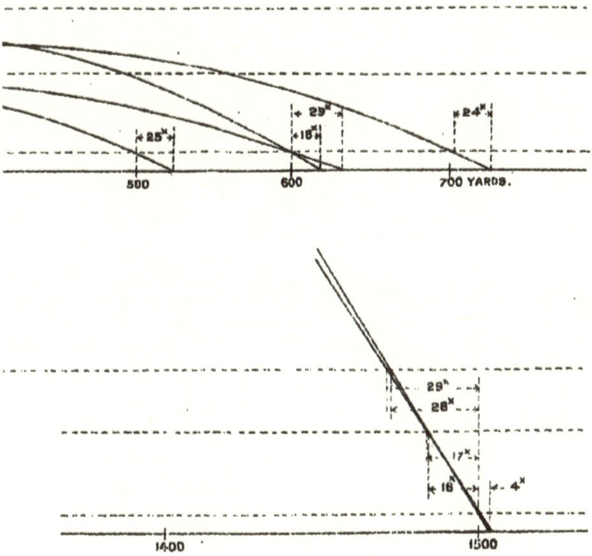

Malby & Sons, Lith.

ii. The height of the object fired at.
iii. The flatness of the trajectory.
iv. The conformation of the ground.

In regard to :—

i. The nearer the rifle is to the ground ii. The higher the object fired at ... ... iii. The flatter the trajectory ... ... iv. The more nearly the slope of the ground conforms to the angle at which the bullet falls	The greater is the dangerous space.

166. A general knowledge of the angle of fall of the bullet in the last 100 yards of its flight, at the shorter ranges, is essential as a guide in deciding when individual fire may be opened with effect (*see* Trajectory Table, page 67). As has been stated, the longer the range the more abruptly does the bullet fall ; consequently, the greater the distance the more accurately must the range be ascertained. Hence, the limits of individual fire are to a great extent governed by the curve of the trajectory and the power of correctly estimating ranges ; and unless the strike of the bullet can be observed, individual fire cannot be effective on small targets at the longer ranges.

167. The firer must also thoroughly understand how the dangerous space is affected by the conditions mentioned in para. 165.

E.g. Suppose that he is firing in the prone position with Mark VI ammunition, and aims at the ground line of a prone figure at 500 yards range, the dangerous space of his fire will be about 50 yards. If, however, he stands to fire, the dangerous space is reduced to about 40 yards.

Section 27.—*Ricochets.*

168. Bullets which rebound after striking the ground or any other obstacle and continue their flight, are said to ricochet.

Ricochets may occur from any surface, and bullets may ricochet two or even three times before their flight is finally arrested. At long range, they are less likely to ricochet from soft ground than from hard smooth surfaces.

Height of Trajectory (in feet) above the Line of Sight of the S.M.L.E. Mark III Rifle.

(*a*) Mark VI ammunition = M.V. 2,060 f.s.

(*b*) Mark VII ammunition = M.V. 2,440 f.s.

(*a*) *Mark VI.*

Range in Yards.	200	300	400	500	600	700	800	900	1,000	1,100	1,200	1,300	1,400
200													
300	1·0												
400	2·1	1·7											
500	3·4	3·7	2·6										
600	4·9	5·9	5·6	3·7									
700	6·5	8·4	8·9	7·8	5·0								
800	8·4	11·1	12·8	12·4	10·5	6·4							
900	10·5	14·3	16·8	17·7	16·8	13·7	8·4						
1,000	12·8	17·8	21·5	23·6	23·8	22·0	17·8	10·6					
1,100	15·4	21·7	26·7	30·0	31·7	31·2	28·4	22·4	13·1				
1,200	18·3	26·0	32·5	37·3	40·3	41·2	39·8	35·3	27·5	15·8			
1,300	21	31	39	45	50	52	52	49	43	33	19		
1,400	25	36	46	54	60	64	66	65	60	52	39	22	
1,500	29	41	53	63	71	77	81	82	79	72	62	46	26
1,600	33	48	62	73	84	92	98	100	99	95	86	73	55
1,700	37	54	70	84	97	107	115	120	122	119	113	102	86
1,800	42	62	80	97	111	124	134	142	146	146	142	133	120
1,900	47	69	90	109	127	142	155	165	172	174	173	167	155
2,000	53	78	101	123	143	161	177	190	199	205	206	203	194
2,100	59	86	113	138	161	182	201	216	229	237	241	241	235
2,200	65	96	125	154	180	204	226	215	260	271	278	282	279
2,300	72	106	139	170	200	227	252	274	293	307	318	324	325
2,400	79	116	153	188	221	250	280	306	328	346	360	369	374
2,500	86	128	168	207	243	278	311	340	366	387	405	418	427
2,600	94	140	184	227	268	307	343	376	406	432	454	471	483
2,700	103	153	202	249	295	338	379	416	450	481	507	528	545
2,800	115	163	223	274	324	372	418	460	499	534	566	592	612

(a) *Mark VI—continued.*

Range in Yards.	1,500	1,600	1,700	1,800	1,900	2,000	2,100	2,200	2,300	2,400	2,500	2,600	2,700
1,600	31												
1,700	64	36											
1,800	100	74	41										
1,900	139	115	85	46									
2,000	180	159	131	96	52								
2,100	224	206	181	148	108	58							
2,200	270	256	234	204	167	120	65						
2,300	320	303	289	263	229	186	134	72					
2,400	372	361	348	326	295	255	206	148	79				
2,500	429	424	412	393	366	330	285	230	165	90			
2,600	489	488	481	465	442	410	369	318	257	185	99		
2,700	555	558	555	545	522	497	461	414	357	290	208	113	
2,800	627	636	637	631	616	594	561	519	467	405	327	237	128

(b) *Mark VII.*

Range in Yards.	200	300	400	500	600	700	800	900	1,000	1,100	1,200	1,300	1,400
200													
300	·6												
400	1·3	1·0											
500	2·1	2·3	1·7										
600	3·1	3·8	3·7	2·5									
700	4·3	5·6	6·1	5·5	3·6								
800	5·7	7·6	8·8	8·9	7·6	4·8							
900	7·3	10·0	12·0	12·8	12·4	10·3	6·3						
1,000	9·1	12·7	15·5	17·5	17·8	16·8	13·5	8·2					
1,100	11·2	15·8	19·7	22·7	24·1	24·1	22·0	17·6	10·5				
1,200	13·5	19·4	24·4	28·6	31·1	32·4	31·4	28·3	22·2	12·9			
1,300	16	22	30	35	39	42	42	40	36	28	16		
1,400	19	28	36	43	48	53	54	54	51	45	35	20	
1,500	23	33	43	52	59	64	68	69	68	63	55	42	24
1,600	26	39	50	61	70	78	83	86	87	84	77	66	50
1,700	31	45	59	71	82	92	99	105	108	107	103	94	79
1,800	35	52	66	83	96	108	118	126	131	132	130	124	112
1,900	40	60	73	96	112	127	139	149	157	161	161	157	148
2,000	46	68	90	110	129	146	162	175	185	192	195	194	188
2,100	52	78	102	125	147	168	186	202	216	226	232	234	230
2,200	59	88	116	143	164	192	214	233	250	264	273	278	278
2,300	67	99	131	161	190	218	243	267	287	305	318	326	330
2,400	75	111	147	181	215	246	276	305	328	349	366	379	386
2,500	84	124	164	204	241	277	311	343	372	397	419	436	443
2,600	93	139	184	228	270	311	349	386	420	450	477	498	515
2,700	104	155	205	254	303	348	392	433	472	507	540	566	588
2,800	115	172	228	282	339	378	437	484	529	570	607	640	667

(*b*) *Mark VII—continued.*

Range in Yards.	1,500	1,600	1,700	1,800	1,900	2,000	2,100	2,200	2,300	2,400	2,500	2,600	2,700
1,600	28												
1,700	59	33											
1,800	94	70	39										
1,900	133	112	83	46									
2,000	175	157	131	97	54								
2,100	221	206	183	152	112	62							
2,200	272	261	241	214	177	129	71						
2,300	328	319	304	280	246	203	148	81					
2,400	388	384	372	353	323	283	233	169	92				
2,500	459	454	447	433	406	370	324	265	193	105			
2,600	526	531	528	517	496	465	424	369	301	218	118		
2,700	604	613	616	609	594	568	532	482	419	340	245	133	
2,800	689	704	711	711	700	680	649	604	547	474	384	277	149

NOTE.—(a) The trajectories have been calculated, using the formula:—

$h \text{ ft.} = 3 X \tan(a_R - a_X)$ where

X = the distance to the point at which the ordinate is required.

a_R = the elevation for the whole range of R yards.

a_X = the elevation for a range of X yards.

(b) The trajectories of the other patterns of rifle may be taken as practically identical with the above.

Range Table of the Short Magazine Lee-Enfield Rifle, Mark III.

For (*a*) Mark VI ammunition, giving a muzzle velocity of 2,060 f.s. (·03 inch foresight).

(*b*) Mark VII ammunition, giving a muzzle velocity of 2,440 f.s. (·00 inch foresight).

Range.	*Angle of Tangent Elevation on Rifle.		†Angle of Descent for the last 100 yards of each range.	
	Mark VI.	Mark VII.	Mark VI.	Mark VII.
Yards.	° ′	° ′	° ′	° ′
200	0 2·0	0 8·0	—	—
300	0 7·5	0 11·3	—	—
400	0 14·0	0 15·3	0 23·0	0 12·0
500	0 21·5	0 20·2	0 36·0	0 18·0
600	0 30·0	0 25·9	0 51·0	0 27·0
700	0 39·5	0 32·7	1 8·0	0 39·0
800	0 50·0	0 40·5	1 27·0	0 54·0
900	1 2·0	0 49·6	1 48·0	1 12·0
1,000	1 15·5	1 0·1	2 12·0	1 33·0
1,100	1 30·5	1 12·1	2 40·0	1 59·0
1,200	1 47·0	1 25·7	3 13·0	2 31·0
1,300	2 5·0	1 41·1	3 51·0	3 8·0
1,400	2 24·5	1 58·4	4 34·0	3 50·0
1,500	2 46·0	2 17·8	5 20·0	4 37·0
1,600	3 9·5	2 39·5	6 8·0	5 29·0

* To get the "angle of departure' the "jump" must be added, which has been found experimentally to be between 7 and 8 minutes positive with Mark VI ammunition, and between 4 and 5 minutes negative with Mark VII ammunition.

† Taken from the smoothed curve of angles given by the height of trajectory at 100 yards from the end of each range.

Range.	*Angle of Tangent Elevation on Rifle.		†Angle of Descent for the last 100 yards of each range.	
	Mark VI.	Mark VII.	Mark VI.	Mark VII.
Yards.	° ′	° ′	° ′	° ′
1,700	3 35·0	3 3·5	7 0·0	6 26·0
1,800	4 2·5	3 30·1	7 58·0	7 29·0
1,900	4 32·0	3 59·5	9 0·0	8 39·0
2,000	5 3·5	4 31·8	10 4·0	9 57·0
2,100	5 37·0	5 7·2	11 15·0	11 24·0
2,200	6 12·5	5 45·9	12 33·0	13 1·0
2,300	6 50·0	6 28·2	13 59·0	14 48·0
2,400	7 29·5	7 14·2	15 33·0	16 45·0
2,500	8 12·5	8 4·1	17 16·0	18 53·0
2,600	8 58·0	8 58·2	19 8·0	21 12·0
2,700	9 48·0	9 56·7	21 10·0	23 42·0
2,800	10 42·5	10 59·8	23 23·0	26 24·0

* To get the "angle of departure" the "jump" must be added, which has been found experimentally to be between 7 and 8 minutes positive with Mark VI ammunition, and between 4 and 5 minutes negative with Mark VII ammunition.

† Taken from the smoothed curve of angles given by the height of trajectory at 100 yards from the end of each range.

The extreme range of the rifle may be taken as 3,760 yards with Mark VI ammunition. This was obtained with a strong rear wind. With Mark VII ammunition the extreme range is from 200-300 yards less.

Section 28.—Firing up and down hill.

169. When a shot is fired at a target placed on the same level as the firer, the forces acting on the bullet cause it to travel in its greatest curve, and the greatest elevation for any given distance must therefore be given to the rifle. If a shot is fired perpendicularly upwards or downwards no elevation is required, for the bullet will travel in an approximately straight line until its impetus is exhausted. Hence it follows that when shooting up or down hill less elevation is necessary than when the object is on the same level. The elevation to be used can be best ascertained by careful observation of fire.

Section 29.—Barometric pressure and temperature.

170. Rifles are sighted for the following conditions:—

i. Barometric pressure, 30 inches (sea-level).
ii. Thermometer, 60° Fahrenheit.
iii. Still air.
iv. A horizontal line of sight.

171. When the barometer rises above 30 inches, more elevation than that normally required for the distance will be necessary, owing to the greater resistance offered to the bullet by dense atmosphere. If the barometer falls below 30 inches, as is the case in damp weather or at a height above sea-level, less elevation will be required than that marked on the sights for the distance, as the atmosphere will offer reduced resistance to the bullet. In the same manner the bullet meets with less resistance in hot weather when the thermometer is high, and greater resistance in cold weather when it is low.

172. The following rule for correction in case of variations in barometric pressure is approximately correct:—

For every inch the barometer rises or falls, add, or deduct, 1½ yards per 100 yards of range.

Thus a reduction of some 30 yards in 2,000 yards range would be required if the barometer stood at 29 inches. The barometer falls about 1 inch for every 1,000 feet of altitude. Therefore, at an altitude of 5,000 feet, it would, in normal conditions of weather, stand at 25 inches, and for 2,000 yards range the elevation required would be 1,850 yards only. Every degree which the temperature rises or falls above or below 60° necessitates the subtraction or addition of about one-tenth of a yard for each hundred yards of range. For example, with the barometer standing at 30 inches, the temperature 100°, and the distance of the target ascertained to be 1,250 yards, the true sighting will be 1,200 yards. Or if the barometric pressure is normal, the temperature 20°, and the distance 1,250 yards, the true sighting will be 1,300 yards. At an altitude of 5,000 feet, given average weather, temperature 100°, and the ascertained distance 1,250 yards, the necessary subtraction to correct for altitude would be 100 yards roughly, and to allow for increased temperature 50 yards. Hence the correct elevation would be 1,100 yards. On the other hand, at an altitude of 5,000 feet in similar weather, with a temperature of 20°, the correct elevation for the same range would be 1,200 yards.

Section 30.—*Wind and light.*

173. The chief cause of trouble to the firer in individual firing is the effect of wind on the path of the bullet. The direction and strength of the wind can be judged by watching trees, grass, &c., by observing the rate of movement of the radiation of heat from the ground (commonly called "mirage"), and by personal sensation.

174. A side wind acts on the greater surface of the bullet and has consequently more influence on its flight than a wind blowing from the front or rear. A wind from the front retards it, and demands more elevation; one from the rear lessens the resistance of the air, and therefore calls for less elevation. No rules can be framed to guide the firer: he should study the effect of wind blowing 10, 20, and 30 miles per hour at one or more distances, such as 500 and 1,000 yards, and having become familiar with the allowances required, he should make proportionate corrections for intermediate winds—oblique winds and winds at intermediate distances. But it must be remembered that, owing to the increased time during which the bullet is exposed to its effect, and to the height attained in its flight, the allowance for wind at long range is out of all proportion to that necessary at short range.

175. In bad light the foresight is less distinctly seen than in good light, and more of it is unconsciously taken into the line of sight. This naturally affects the elevation used, less being required on a dull than on a bright day.

Section 31.—*Use of collective fire.*

176. From the many causes which affect marksmanship, such as the condition of the firer, the atmosphere, heat, light, error of the rifle, imperfection of ammunition, uncertainty in estimation of range, the difficulty of aiming at a small indistinct or invisible objective, the steepness of the fall of the bullet rapidly accentuated as the range increases, &c., it is evident that, under service conditions, at the longer ranges, effect must be looked for from collective fire rather than from individual effort.

Section 32.—The dispersion of individual and collective fire.

177. Owing to imperfections in the rifle and ammunition and also to errors on the part of the firer, it is found that a series of shots fired at a large target by an individual do not all strike the point for which the sights are set, but they form a group of shot marks the density of which varies mainly with the skill of the firer. Plate XII shows a group of shot marks divided equally into 8 vertical and 8 horizontal bands in such a way as to represent the theoretical distribution per cent. of a number of shots fired without alteration of sighting. It is evident that the trajectories of these shots will not coincide, but form a figure termed the "Cone of Fire" (*see* Plate XIII). If in this way the mean errors in a horizontal and vertical sense of a number of individuals be determined by experiment, a reliable guide will be established as to the greatest distances at which individual fire can be expected to be effective against targets of varying size (*see* Plate XIV) which are clearly visible to the unassisted eye. The diagrams on Plate XIV represent the grouping diagrams of expert shots.

178. It is clear that, when the aim is well directed, the bullets should always strike an object so long as the grouping is either smaller or of the same size as the object. But when the group is spread over a larger surface, shots must necessarily miss the mark in proportion as the size of the grouping exceeds that of the objective.

179. When a body of soldiers fires with the same elevation at the same object, the dispersion is accentuated by the varying skill and eyesight of the men, with the result that the cone formed will be of larger dimensions than in the

PLATE XII.

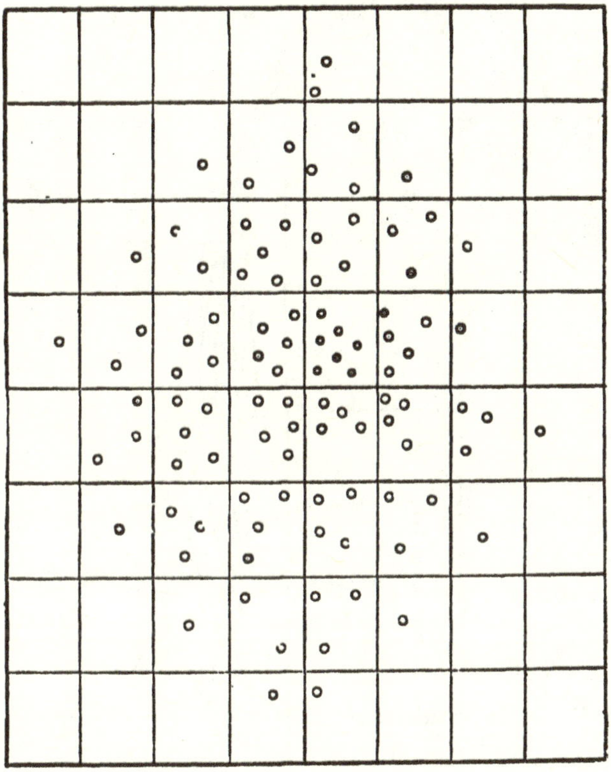

PLATE XIII.

CONE OF FIRE.

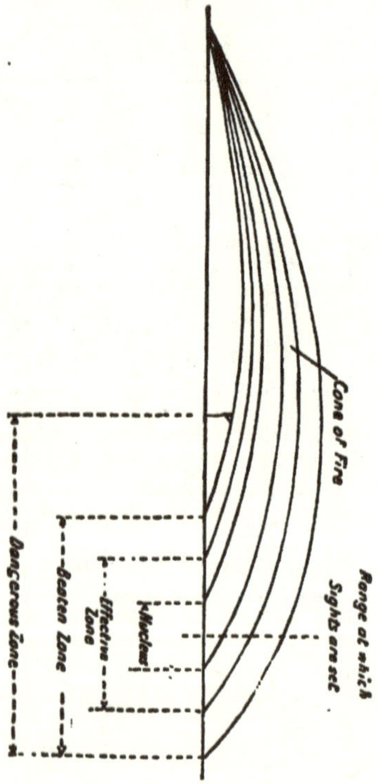

PLATE XIV.

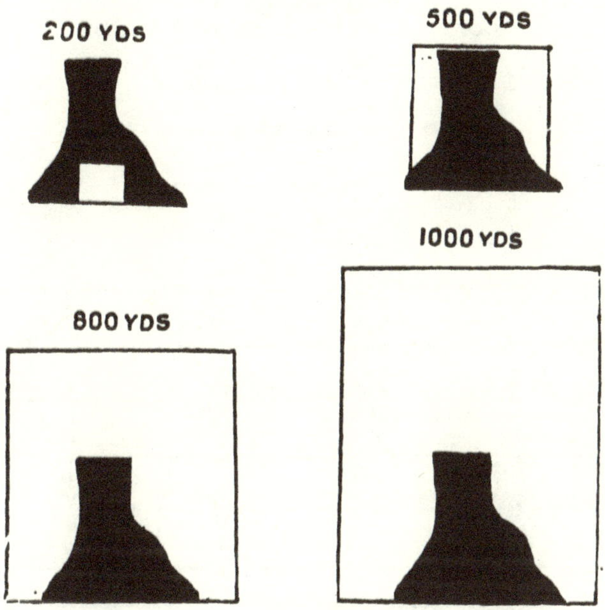

SCALE $\frac{1}{24}$.

Diagrams showing rectangles containing shot groups fired in the prone position with wrist and rifle unsupported, compared with a figure target having an area of 200 sq. inches.

PLATE XV.

DEPTH DISPERSION OF CONCENTRATED COLLECTIVE FIRE AT 1,500 YARDS.

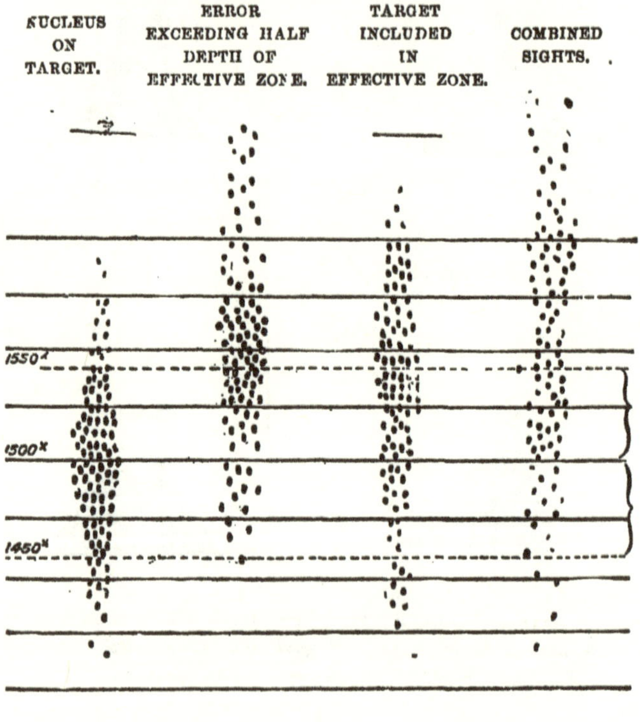

PLATE XVI.

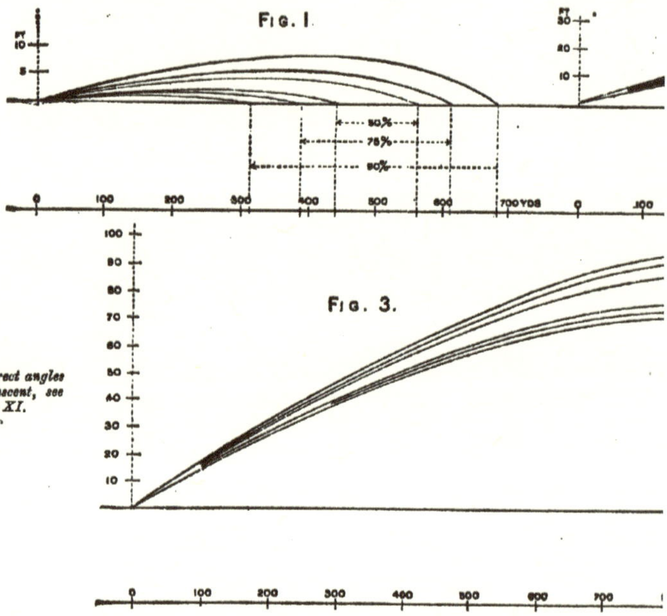

For correct angles of descent, see Plate XI.

FIG. 2.

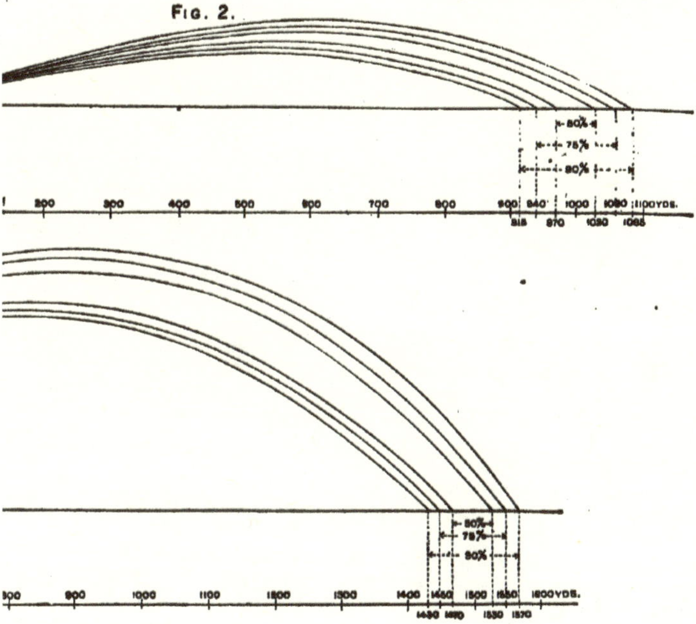
50%
75%
90%
200
300
400
500
700
800
900
940
1000
1100 YDS.
970
1065
50%
75%
90%
900
1000
1100
1200
1300
1400
1450
1500
1600 YDS.
1430
1530
1570

case of individual fire. The dimensions will also be further increased if the firers are, from any cause, unsteady, the aiming target indistinctly seen, the rifles in bad condition, &c.

180. The area of ground beaten by a cone of fire is termed the "Beaten Zone." It is regarded as a plane surface. Plate XIII illustrates the additional space which would have to be taken into account in consequence of the height of the objective.

181. If the results of the shots fired by a body of men are recorded either on a vertical or horizontal surface (*see* Plate XV), it will be found that they are grouped in such a way that the majority of the bullets fall in the general direction of the line of fire, that the density of the grouping decreases progressively from the centre to the extreme limits of the beaten zone, and that the bullets are collected most thickly near the point for which the sights were set. This dense grouping is usually termed the nucleus of the cone of fire, and is regarded for purposes of comparison as including the best 50 shots per cent. (*see* Plate XVI, figs. 1, 2, and 3). The area of ground beaten by the best 75 per cent. of shots fired is termed the "Zone of Effective Fire," for it has been found by experiment that useful results can only be looked for when the target is within these limits.

182. The depth of the beaten zone at short ranges need not be considered on account of the flatness of the trajectory, which ensures that practically the whole extent from the firers to the target is swept by bullets. (*See* para. **161.**)

183. Up to ranges of 1,500 yards, when the ground is parallel to the line of sight, the depth of the beaten zone decreases with the range on account of the increased angle of descent of the bullets. At longer ranges it becomes deeper again, in consequence of the influence of atmospheric conditions on the flight of the bullet, the increased effects of errors in aiming, and the faults in the rifle and ammunition. On the other hand, owing to faults in aiming, comparative invisibility of the target, inaccuracies in the rifle and ammunition, and atmospheric influences, the lateral dispersion of a cone of fire increases as the range becomes greater. With Mark VI ammunition the dispersion of the best 75 per cent. of shots fired may be taken as 7 feet by 220 yards at 500 yards, 14 feet by 120 yards at 1,000 yards, and 22 feet by 100 yards at 1,500 yards.

184. In spite of the fact that up to 1,500 yards the depth of the beaten zone is diminished, fire is less effective than at short distances, owing to the increased steepness of the angle of descent, and this becomes more marked beyond that range.

Section 33.—*Ground in relation to fire effect.*

185. So important is the influence exercised on fire effect by the shape of the ground in relation to the grouping of bullets, that it is essential for all officers and non-commissioned officers to understand thoroughly how the probability of fire effect is increased or diminished by the inclination of ground with reference to the trajectory.

In attack, such knowledge will assist them in adopting formations and directing the fire of their men to the best advantage; in defence it will aid them to select the best positions for fire action, and will enable them to take steps to minimize their inherent disadvantages. In the following examples only plane surfaces, without undulations or accidents, are considered.

PLATE XVII.

Ground rising in respect of line of sight.

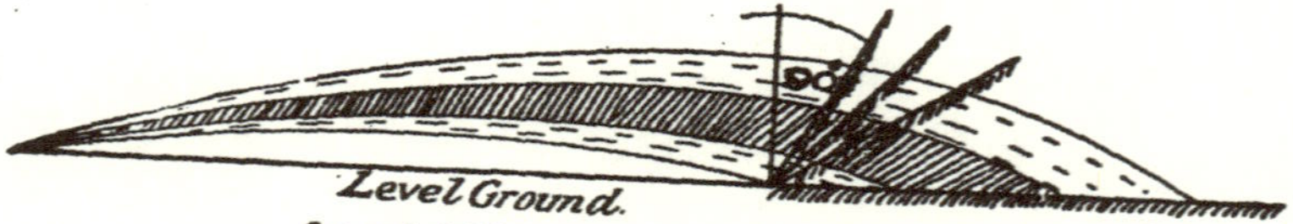

Ground falling in respect of line of sight.

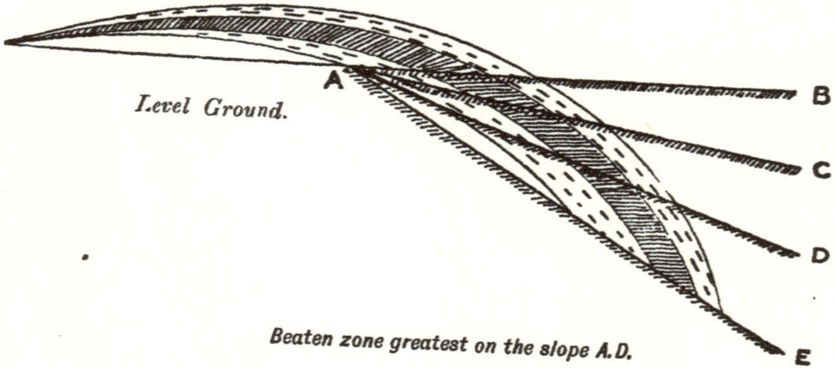

Beaten zone greatest on the slope A.D.

186. On level ground, the zone beaten by collective rifle fire varies considerably with the range (*see* para. **183**). Its extent is further influenced by the inclination of the ground to the line of sight.

187. When the ground rises with reference to the line of sight, the depth of the zone beaten by bullets is decreased, and is least when the angle between the ground and the trajectory is 90°. *e.g.*, with Mark VI ammunition, when firing at ground rising at 2°, 5°, and 10°, the depth of the beaten zone at 1,500 yards range is decreased roughly by $\frac{1}{4}$, $\frac{1}{2}$, and $\frac{2}{3}$ respectively. (*See* Plate XVII.)

188. Since the grouping of the bullets becomes closer as the upward slope of the ground increases, the effect of errors in estimation of range will be more serious, and as the bullets will fall at a steep angle the dangerous space will be proportionately reduced. On such ground, therefore, troops should be drawn up in shallow formations, but supports and reserves may be nearer the firing line than is normally advisable.

189. When the ground beaten by bullets falls in respect of the line of sight, the depth of the beaten zone is augmented, in proportion as the downward slope increases, until it reaches its greatest magnitude when the angle of the fall of the bullets is the same as the slope of the ground; or, in other words, when the trajectory is practically parallel to the ground surface. (*See* Plate XVII.) In these circumstances, the fire becomes grazing, and the extent of the dangerous space is nearly identical with the beaten zone. Therefore, at short ranges, where the trajectories are flat, the depth of the beaten zone will be much increased if the ground behind the target falls at a gentle slope. At long ranges, on the other hand, a greater area will be beaten when ground falls more steeply, *e.g.*, at 1,500 yards range, the depth of the beaten zone is roughly increased by $\frac{3}{5}$ when the bullets strike ground falling at 2°. On ground falling at 5°, which is nearly parallel to the trajectory at this range, the depth beaten is about ten times greater than on level ground.

190. It is clear, therefore, that ground far behind the objective will at times be swept by unaimed fire, and it follows that in such circumstances supports should either be under cover, or, if there is no cover, in shallow columns on narrow frontages with the object of reducing target surface as much as possible.

191. When the objective is a crest line, the depth of the beaten zone is greatest, and part of the fire is grazing, when the ground beyond the crest is parallel, or nearly so, to the trajectory of the bullets. In this case, at short ranges there will be behind the crest line a defiladed zone, or space not swept by fire, greater or less according to the distance from which fire is delivered, the inclination of the line of sight, the extent of the hill top, and the inclination of the reverse slope. (*See* Plate XVIII.) It appears, therefore, that when the firing line is placed on the crest of a razor-backed hill with steep reverse slopes, supports and reserves will at all ranges be but little exposed to unaimed fire when posted in its vicinity. In other cases, when the crest of a hill is occupied, the vulnerability of supports and reserves will be least if, when the enemy is at long range, they are withdrawn from, and as he approaches closed on, the firing line.

PLATE XVIII

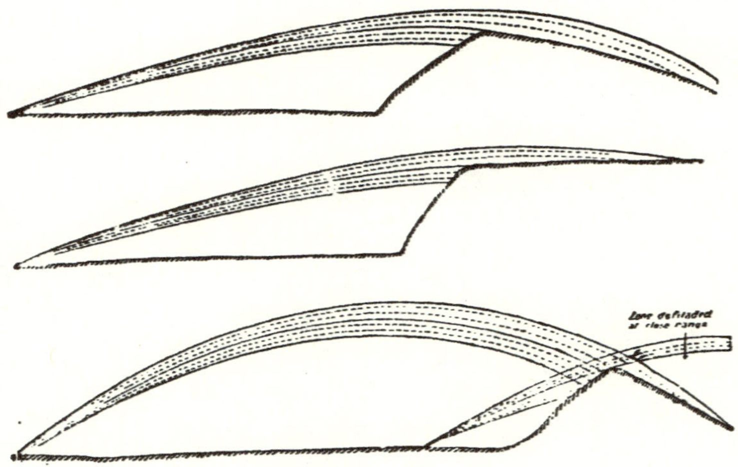

Section 34.—Searching.

192. Fire will not as a rule produce results commensurate with the amount of ammunition expended, or fulfil the purpose for which it is used, unless the target is included within the area beaten by 75 per cent. of the bullets directed upon it in collective fire. When collective fire is concentrated and steadily delivered and the target is clearly seen, the depth of this area of ground will not greatly exceed that of the normal zone of effective fire.

193. If an error in sighting is made which will cause the nucleus of the cone to strike at a distance short of, or beyond the target, equal to half the depth measurement of the effective zone, the target will not be included in this zone, and the fire will be ineffective. (*See* Plate XV.)

194. The depth measurement of the zone of ground beaten by 75 per cent. of the bullets decreases as the range increases up to 1,500 yards; the probability of error in ranging and judging the error of the day increases with the range. That is, the probability of sighting errors increases at the same time as the permissible error in sighting decreases.

195. Under active service conditions it may be assumed that even if the range is measured with a range finder the probability of error in ranging and judging atmospheric influences is such that, at distances beyond 1,000 yards, collective fire concentrated on any target with one sighting will probably be ineffective. To give a satisfactory degree of assurance of fire effect, it is advisable in such cases, unless sighting can be corrected by observation of results, to distribute fire in depth by using two elevations differing by 100 yards. One of these elevations would be 50 yards over and the other 50 yards under the sighting believed to be correct. There will thus be two cones of fire and two beaten rectangles overlapping between the nuclei, and an even distribution of fire will be obtained over a zone about 150 yards deep; the permissible error in sighting will be increased, but the greatest density of the grouping will be reduced. This method of searching or distributing fire in depth is called "Combined Sights" (*see* Plate XV). It may sometimes be used to increase assurance of effect when the enemy's position is ill-defined, but should never be used if observation of results can be obtained.

196. Ordinarily, fire should be closely concentrated with a view to observation of results, but if observation has failed, or if the situation calls for immediate application of effective fire, combined sights should be used at ranges beyond 1,000 yards. Combined sights should not be employed by bodies of less than two platoons.

Section 35.—Night firing.

197. Men should be exercised in the following automatic method of aligning their rifles for night firing at ranges of about 300 yards and under.

Aiming marks should be selected just above the ground line and within 100 yards of the squad. The men should then be ordered to bring their rifles into the firing position with both eyes shut. The right eye should then be opened, and the approximate alignment of the rifle verified. After some practice each man will be able to ascertain his individual tendency, which he should correct with practice until able to align the rifle with his eyes shut with approximate accuracy.

This exercise should be practised in the day-time until proficiency is attained, when men should be practically exercised in firing a few rounds after dark at large screens at a range not exceeding 300 yards, to demonstrate the value of the instruction afforded, the position of the screen being indicated by some rough expedient to represent the flash of a rifle.

Much material effect is not to be anticipated except against an enemy in movement, but the moral effect should be considerable.

A body of troops in a position commanding open ground or an approach which may be used by the enemy, may arrange to sweep it with fire by laying rifles in rests constructed by daylight, by preparing illuminated aiming marks giving a horizontal line of sight, or by firing at the flashes of the enemy's rifles. For details as to the arrangement of rifle rests and as to the illumination of the foreground, *see* Manual of Field Engineering.

Firing at night should only be allowed within quite close range.

CHAPTER IV.

MUSKETRY EXERCISES.

AIMING INSTRUCTION.

Section 36.—Stages of instruction.

198. Aiming will be taught in stages, as follows:—

Explanation of the theories of sighting, elevation, and deflection.

Method of adjusting the sights.

Explanation of aiming with the backsight and foresight.

Aiming with scrupulous accuracy from a rest.

Demonstration of results of inaccuracy in aiming.

Aiming combined with trigger pressing.

Declaring point of aim at moment of discharge.

Aiming from a rest at figures and ground.

Aiming at indistinct targets described by word of mouth.

Aiming off, making allowance for deflection as ordered.

Setting of the windgauge and fine adjustment according to orders.

Judging deflection allowance for wind and aiming off without orders.

Aiming up and down.

Rapid adjustment of sights in accordance with orders.

Aiming and snapping without a rest in all positions.

Rapid alignment of sights.

Aiming with long range sights.

Aiming off for movement.

Fire discipline exercises.

It is immaterial whether the instruction is given in the above order, but further training should be deferred until the eye is shown to be accurate in aiming. Ample time should be devoted to trigger pressing and declaring the point of aim on discharge; this instruction may conveniently be given indoors.

199. Aiming instruction should proceed simultaneously with firing instruction and muscle exercises, short lectures and questions being included in the daily programme.

Section 37.—*Accuracy in aiming.*

200. Especial care will be taken that the sights of any rifle used in aiming instruction are in perfect order.

201. The aiming mark in the earlier stages will be a special aiming target* placed at a distance not exceeding 100 yards, but when the habit of accuracy has been acquired service targets will be used and care will be taken that there is no falling off in accuracy owing to the shape and comparative invisibility of these targets.

202. The instructor must inspect the sights frequently to ensure that they are correctly adjusted in accordance with his orders.

He will explain the following rules, and demonstrate the results to be anticipated from common errors in aiming:—

i. The backsight must be kept upright.

ii. The left or right eye, according to the shoulder from which the man shoots, must be closed.

iii. Aim must be taken by aligning the sights on the centre of the lowest part of the mark, the top of the fore-sight being in the centre of, and in line with, the shoulders of the U or V of the backsight.

---

* For description of target *see* Part II of these Regulations

203. When these principles have been mastered, the instructor will loosen the sling, adjust the sights for any given range and aim from the rest at the target, taking care that his eye is immediately above the butt-plate. It will be convenient to use a sandbag aiming rest to steady the head during the aiming. Having aimed, he will call on each individual to observe the correct method of aligning the sights on a mark. Each recruit will then act similarly, when the instructor will verify the aim, point out errors, and explain how they would have affected the accuracy of the shot, and how they are to be avoided. He should occasionally call on a man to point out any errors which may have been made by his comrades.

204. Extreme accuracy of aim must be insisted on even during the first lesson.

205. In aiming with the long range sights a similar procedure to that which is described above will be pursued, but the eye will be placed about 1 inch behind the small of the butt and the top of the head of the dial sight will be seen in the centre of the aperture.

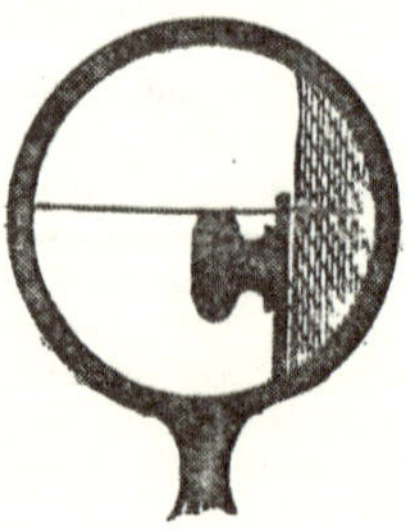

Unless special care is taken, lateral error will be made in centring the head of the dial sight.

Section 38.—Common faults in aiming.

206. The most common faults in aiming are :—

i. Taking too much or too little foresight into the U or V of the backsight.

It should be explained that a fine or half sight will cause the bullet to strike with Mark VI about 5 inches and 3 inches lower respectively, and with Mark VII 7 inches and 4 inches lower respectively per 100 yards of range, than when the correct sight is taken.

The following method will be found useful to guide a recruit in taking the correct amount of foresight :—Lay the edge of a piece of paper on the upper edge of the backsight cap, when the sight will appear as shown in one of the subjoined diagrams. A piece of cardboard laid on the cap and held in place by an elastic band will answer the same purpose.

ii. Inaccurate centring of the foresight in the notch of the backsight.

The recruit should understand that this inaccuracy will deflect the muzzle of the rifle to the side on which the line of aim is taken, *e.g.*, if aim be taken over the right edge of the notch the direction of the line of fire will be to the right of the line of sight.

iii. Fixing the eyes on the foresight, and not on the object. If the eye is accommodated for the foresight, the firer will retain only a blurred image of the target. This may not affect the result of his shooting at stationary vertical targets, but when firing at service targets with appear and reappear, or at areas of ground, it will be necessary for him to watch the target closely.

iv. Inclining the backsight to one side.
In this case the bullet will strike low, and to the side of which the sights are inclined. The resulting error will be considerable at long ranges.

Section 39.—Triangle of error.

207. To test proficiency in aiming and to demonstrate the errors which will arise from inexactitude, the following method of recording a " triangle of error " will be employed.

The rifle will be placed on an aiming rest, and directed at a sheet of blank white paper, affixed to any convenient object at a distance of about 10 yards from the rifle ; a second aiming rest should be used to steady the head in aiming. A marker will stand at one side of the target with a small rod bearing an aiming disc of metal or cardboard, painted white, about 1½ inches in diameter ; on this is a black bull's-eye ½-inch in diameter, in the centre of which is pierced a hole just large enough to admit the point of a pencil. The disc will be held on the paper, the instructor will align the sights on it as laid down in para. 202, and its position will be marked with a pencil.

Each man in succession will be called up and ordered to look along the sights, but without touching the rifle, and when he has satisfied himself as to the correctness of the aim, the disc will be removed. It will then be replaced on the paper and moved at his direction until the lower edge of the bull's-eye is brought into the line of sight, when its position will be marked. The operation will be repeated three times, and the points thus fixed will be joined in such a way as to form a triangle. The position of the points in relation to the instructor's aim* will expose any constant error in aiming. Their position in regard to one another will show inconsistency.

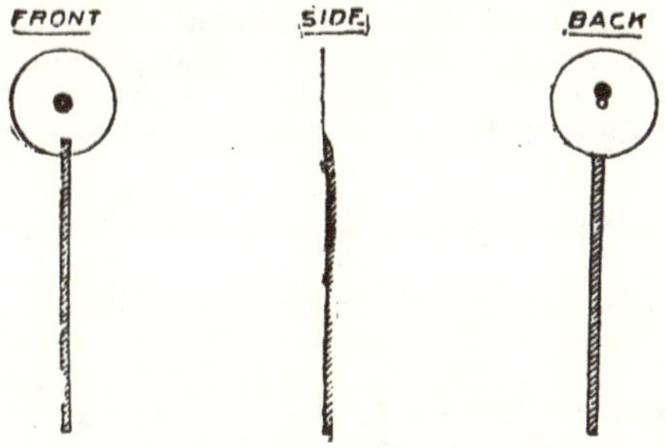

208. If the triangle is formed so that its greatest side lies vertically on the paper, it proves that the soldier's principal fault is inconsistency in respect of the amount of foresight taken up into the line of sight; if horizontally, that his principal error is inaccurate centring of the foresight.

* If the recruit's aim is below that of the instructor, it shows that he has taken too full a sight: if above, that his sighting has been too fine; if to the right that the foresight was on the left of a perpendicular drawn through the centre of the notch; if to the left, *vice versâ*.

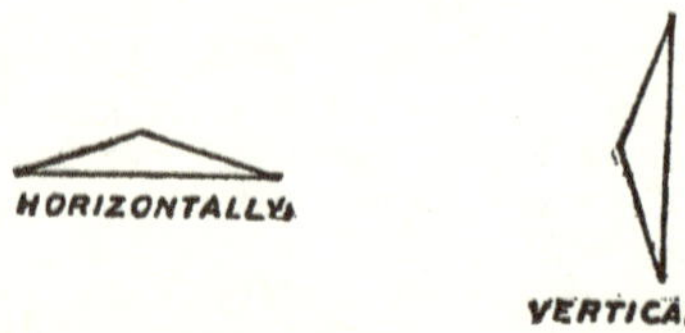

209. When any one of the sides of the triangle exceeds one-third of an inch, or if the centre of the triangle is more than one-third of an inch from the instructor's aim, the recruit will be noted for further instruction. The aim corrector will also be used to enable the instructor to supervise the soldier's methods and test his progress, the aim being taken from an aiming or firing rest.

Section 40.—*Aiming at the ground, and marking down an enemy.*

210. The eyesight must be gradually trained in aiming at figures or other service targets and at ground which might conceal an enemy. For this latter purpose a fatigueman will be directed to show himself for a few seconds at different ranges. The squad will then aim from rests at the ground which they believe to be occupied. After a short interval the fatigueman will stand up, when errors will be corrected and criticized.

211. Further practice will be given in aiming at ill-defined targets described by word of mouth, which will lead up to the fire discipline exercises described in para. 267.

Section 41.—*Aiming off for wind.*

212. The use of the windgauge will be explained, but it is not to be regarded as the normal means of making allowance for wind under service conditions. The instructor, having explained that a variation of one division on the windgauge scale is equivalent to 6 inches on the target per 100 yards of range, will tell the soldier how many feet of deflection allowance would be required for the wind which is blowing,* and then direct him to set his sights for the range, adjust the windgauge, and aim from an aiming rest.

213. The recruit should be taught to discriminate between mild, fresh, and strong winds, by the effect which they exercise on natural objects, and to note the direction of the wind as front, rear, right angle, or oblique, by turning his face full towards the wind.

214. The approximate allowances in elevation or deflection for these winds at decisive range will be taught. Practice will be afforded in aiming off a full-length figure target according to the range, and the strength and direction of the wind, actual or supposed.

215. The recruit should be accustomed to make deflection allowance with reference to the breadth of the target, the full length figure target being rather less than 2 feet wide. Such practices should be limited to 600 yards, but in addition there must be practice in aiming off at all ranges according to orders, the allowance being indicated if possible by reference to the breadth of the target, or intervals in a formation, or by selection of an auxiliary aiming point. If this is not possible, it must be expressed in feet or yards.

---

* A ready method of ascertaining the required adjustment is to multiply the number of feet, or fractions of feet, of allowance for wind by 2, and divide the product by the number of hundreds of yards in the range. The result will be the number of divisions which the windgauge must be moved.

216. In such exercises it is necessary to employ a fatigueman at the target to indicate the correct point of aim with a marking disc after each aim taken by the recruit. The amount of allowance made will be observed by the instructor with an aim corrector, unless an aiming rest is used.

Section 42.—Aiming up and down.

217. In aiming up and down the aim should be directed at a point not more than 3 feet above or below the six o'clock line, according to the position of the target beyond or short of the zone for which the sights are set, but if the difference between the range and the sighting exceeds 200 yards, it will be best to alter the sighting.

218. To practise aiming up and down, fatiguemen or disappearing targets should be brought into view for short periods of time at various ranges, the sights being fixed as may be ordered.

Section 43.—Rapid adjustment of sights.

219. Instructors will frequently test their squads in setting their sights rapidly and accurately. They will give orders for aiming or firing at definite targets, but without naming the range, and require their squads to adjust their sights before they bring the rifle to the shoulder. The careful adjustment of sights will thus become instinctive.

220. Adjustment of sights will also be practised in connection with movement of the firer or the target.

221. When there is time, the sights will be adjusted for every alteration in the range, and aim will be taken at the lowest part of the mark, but frequent small changes of sighting lead as a rule to loss of fire effect, and may be avoided by aiming up or down. (Section 42.)

Section 44.—Aiming off for movement.

222. The instructions for aiming at moving targets in the following paragraphs will be given during the latter portion of preliminary training, practice in this form of shooting being carried out on 30 yards or miniature ranges, where the pace of movement may be regulated in strict accordance with that of service targets.

223. When firing at crossing targets, aim will first be taken on the object, then following it sideways, the aim will be carried in advance and kept in front of the object until the rifle has been fired. The distance to which the aim should be carried in advance of the target will vary according to the range, rate of movement, and direction of the movement.

224. Up to 500 yards range, aim should be taken :—

About 1 ft. in front per 100 yards, at a single man walking.
" 2 " " " at a single man doubling.
" 3 " " " at a single horseman trotting.
" 4 " " " at a single horseman galloping.

225. Thus, at 100 yards, a soldier should aim about the breadth of a man in front of an individual walking, and at 200 yards about a horse-length in front of a single horseman trotting. Fire will rarely be effective at a single man moving across the front at more than 300 yards range, or at a single horseman above 500 yards. At effective and long ranges, aim should be taken at the head of a body of troops moving to a flank.

226. In aiming at an object moving directly or obliquely towards or away from the firer, allowance for elevation must usually be made by aiming lower or higher.

Firing Instruction.

Section 45.—Instructors.

227. Instructors will teach by force of example rather than by word of mouth, and be careful to refrain from any form of comment which may discourage young soldiers. Words of command are seldom required except in collective firing instruction, all motions of firing being performed independently, and each man being required to use his own judgment as much as possible. Faults must not, however, be overlooked or allowed to become formed habits; the essential points of the firing positions are to be insisted upon from the beginning, as the foundation of fire discipline.

228. In the early part of training, squads will not, as a rule, consist of more than seven men, who will be assembled round the instructor in a semicircle. The instructor will explain the uses of the different firing positions and illustrate them to the squad. Recruits will practise the motions separately until able to combine them, and assume each position rapidly and without constraint.

229. The position of each individual will be corrected in turn; the regulation positions may be varied if physical characteristics render them unsuitable in any case, but awkwardness in the first stages of instruction will not be accepted as an indication that the regulation position requires modification.

230. The instructor should stand about 5 paces from the recruit and to his right front while correcting his faults.

*Section 46.—Firing rest.**

231. The firing rest will be frequently employed in early instruction to enable the recruit to support the rifle and rest his muscles, whilst the instructor modifies, or corrects, his position. An incorrect position, however, usually arises from want of accuracy in the preliminary actions which lead to it, and it is to these that attention must be given, for a faulty position once acquired cannot easily be corrected.

Section 47.—Trigger pressing.

232. Before he is permitted to practise snapping, the recruit will be given several lessons in the correct way of pressing the trigger. The rifle will be rested on sandbags or in an aiming rest, and the recruit will be seated with his elbows rested on a table. The instructor will first take steps to ensure that the recruit can move his trigger finger independently of the remainder of the hand and arm.

233. The instructor will explain (i) that as the trigger has a double pull-off two distinct pressures are necessary to fire the rifle (the strength of the first pull is 3 to 4 lbs.; that of the second, 5 to 6 lbs.). The first pull should be taken when the rifle has been brought into the position for aiming; the second, when the sights are aligned on the mark. (*N.B.* The charger-loading Lee-Enfield rifle has a single pull-off; strength of pull, 5 lbs.). (ii) That the direction of the pull-off is diagonally across the small of the butt. (iii) That the first joint of the forefinger should be placed round the *lower* part of the trigger; and (iv) that in order not to disturb the aim breathing must be restrained when pressing the trigger.

---

* For description of rest *see* Part II of these Regulations.

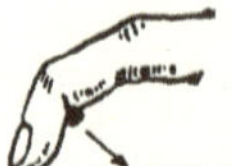

234. In order that he may learn from experience the pressure required to release the cocking-piece, the soldier will also be directed to place his forefinger under that of the instructor, but without exercising pressure, whilst the instructor carries out the motion. Then, to enable the instructor to ascertain whether the method is understood, the soldier will place his finger over that of the instructor, and exert the pressure. Finally, the soldier will himself press the trigger, while the instructor uses the aim corrector. The main object is to release the cocking-piece without disturbing the aim.

235. Special care will be taken that the breathing is restrained while pressing the trigger, and the recruit is always to say after the spring is released whether the aim was maintained truly at the moment of snapping. If not, he must state definitely the direction in which the rifle was pointed at the moment of discharge.

236. From time to time the instructor will test the aim and steadiness of each recruit with an aim corrector, and if necessary further lessons in trigger pressing will be given. Practice with the sub-target or portable sub-target machine or with miniature cartridge may advantageously be given, to develop steadiness until range practice is begun.

237. Trigger pressing requires most careful individual instruction, during which the necessity for determination

and strong personal effort will be impressed on the mind of every recruit.

238. Daily practice in snapping is required by trained soldiers as well as recruits. Progress should be tested by grouping practice on the 30 yards or the miniature cartridge range or with a sub-target machine.

239. The use of the sling for steadying the rifle during firing is not to be taught (*see* para. 457).

Section 48.—*Firing standing.*

240. The standing position will as a rule be used on service to fire from breastworks, high walls, and cover, such as long grass or standing corn, or to take a snap shot, when advancing, so that the pace of the advance is not materially checked.

241. It is a convenient position for elementary instruction, but when recruits have acquired facility in handling the rifle, they will be practised for the most part in the prone position in the open, and occasionally in the kneeling position, or lying behind cover.

242. *To load.*—On the command "*Load,*" turn half right; carry the left foot to the left and slightly forward so that the body is equally balanced on both feet. Bring the rifle to the right side in front of the hip, with the muzzle pointing upwards, small of the butt just in front of the hip, grasping the stock with the left hand immediately in front of the magazine. Turn the safety catch completely over to the front with the thumb or forefinger of the right hand. [*Charger-loading Lee-Enfield rifle.*—Lower the safety catch with the thumb of the right hand.] Pull out the cut-off if closed, first pressing it downwards with the thumb, then seize the knob with the forefinger and thumb of the right hand, turn it sharply upwards, and draw back the bolt to its full extent. Take a charger between the thumb and first two fingers of the right hand and place it vertically in the guides. Then, placing the ball of the thumb immediately in front of the charger, and hooking the forefinger under the cut-off, force the cartridges down with a firm and continuous pressure until the top cartridge has engaged in the magazine.* Force the bolt sharply home, turning the knob well down, and, with the thumb or forefinger of the right hand, turn the safety catch completely over to the rear. [*Charger-loading Lee-Enfield rifle.*—Raise the safety catch with the forefinger of the right hand.] Then button the pouch, seize the rifle with the right hand in front of the left, bring the left foot back to the right and order arms (*see* Plate XIX).†

Note.—The command "*Load*" is only required for drill purposes, or when charging rifles before leaving quarters on service. After the rifle is once charged the soldier is responsible that his magazine is refilled at once, whenever it has been emptied.

243. *To unload.*—As when loading, but after drawing back the bolt, without turning the knob down, work the bolt rapidly backwards and forwards until the cartridges are removed from the magazine and chamber, allowing them to fall on the ground; then close the breech, press the trigger,

* The magazine will hold two charges of five cartridges each, but should, in ordinary circumstances, be loaded with one only, as the soldier will thus retain the power of adding another charge, at any time, should necessity demand. If, when on the line of march, it is desired to charge the magazine without loading the rifle, the top cartridge may be pressed downwards with the thumb and the cut-off closed (*see* para. **264**).

† Following page 104.

close the cut-off by placing the right hand over the bolt and pressing the cut-off inwards, apply the safety catch, lower the backsight or the long range sights, and order arms.

244. *To adjust the backsight.*—Hold the rifle in the loading position so that the lines on the backsight can be clearly seen. Press in the stud (or studs) on the side of the slide with the left or right hand; move the slide until the line is even with the place on the leaf giving the elevation for the distance named, taking care that it is firmly fixed.

[*Charger-loading Lee-Enfield rifle.*—Loosen the clamping screw with the thumb and forefinger of the right hand, move the slide until the top is even with the line on the leaf giving the elevation for the distance named, and then tighten the clamping screw.]

245. *To lower the backsight.*—Press the stud or studs inwards with the left or right hand, and draw the slide backward as far as possible.

[*Charger-loading Lee-Enfield rifle.*—Loosen the clamping screw, lower the slide to the bottom of the leaf with the forefinger and thumb of the right hand, then tighten the clamping screw.]

246. *Adjustment of the windgauge.*—Turn the screw on the right of the windgauge till the central line is level with the line on the scale giving the required deflection.

[*Charger-loading Lee-Enfield rifle.*—Turn the screw on the left of the slide until the line on the windgauge is level with the line on the scale giving the required deflection.]

247. *Fine adjustment. S.M.L.E. rifle, Marks III and IV.*—Press the stud on the slide with the thumb of the left hand till the worm wheel can be easily revolved; turn the worm wheel with the thumb nail of the right hand, until the required elevation is obtained (*see* para. 32). The stud must not be pressed to such an extent that the worm wheel is entirely disengaged from the rack.

[*Other Marks.*—Turn the fine adjustment screw with the forefinger and thumb of the right hand until the line on the leaf is level with the line on the fine adjustment scale giving the required elevation.]

248. *To adjust the long range sights.*—Hold the rifle in the loading position so that the dial can be clearly seen. Move the pointer to the place on the dial giving the elevation for the distance named, and then raise the aperture sight.

249. *To aim and fire.*—Load, and direct the eyes on the mark. Then bring the rifle into the hollow of the right shoulder, press it in with the left hand, grasp the small firmly with the thumb and three fingers of the right hand, place the forefinger round the lower part of the trigger, and exert sufficient pressure to take the first pull [S.M.L.E. Rifles only, *see* paras. 40 and 233]; the backsight to be upright, left elbow well under the rifle, right elbow a little lower than, and well to the front of, the right shoulder; as the rifle touches the shoulder bring the cheek down on the butt, keeping the face well back from the right hand and cocking-piece, close the left eye, align the sights on the mark, restrain the breathing, and press the trigger (*see* Plates XX and XXI).* After a pause bring the rifle to the loading position and repeat the practice, or apply the safety catch and order arms.

250. Great care must be exercised to ensure that the forefinger is not placed on the trigger before the rifle is in contact with the shoulder, and that a firm grip is maintained with both hands while firing.

---

* Following page 104.

251. During this exercise the instructor will explain to the recruits that uniformly accurate shooting cannot be obtained unless the butt rests firmly in the hollow of the shoulder, and that the further the eye is kept from the backsight the more clearly will the sights be defined, the less strained the position of the head and neck, and the less the effect of recoil.

252. The firer should always declare the direction of his aim at the moment of discharge before removing the rifle from the shoulder.

Section 49.—Firing prone.

253. The prone position will generally be adopted by troops on open ground, or when firing from continuous low cover, or from behind small rocks, trees, ant-heaps, &c. Preliminary instruction in firing, and firing exercises, will as a rule be carried out in the prone position.

254. i. *To lie down.*—Turn half right, bring the rifle to the right side as when standing. Place the right hand on the ground, and lie down on the stomach obliquely to the line of fire, with the legs separated, left shoulder well forward, left arm extended to the front, and rifle resting on the ground in a convenient position, muzzle pointing to the front.

ii. *To load.*—As when standing.

iii. *To unload.*—As when standing.

iv. *To adjust sights.*—Draw the rifle back with the left hand [*S.M.L.E. and charger-loading Lee-Enfield rifle*—draw the rifle back through the left hand] until the lines on the backsight can be clearly seen, and proceed as when standing.

v. *To aim and fire.*—Proceed as when standing. To obtain elevation, the body must be raised on the elbows and slightly retired. (*See* Plates XXII and XXIII.)*

255. Recruits will be trained to assume the prone position rapidly, and to perform the loading and aiming motions with as little movement as possible. The oblique angle is not to be unnecessarily exaggerated. Behind objects affording a restricted amount of cover, the body and legs will be parallel to the line of fire, if the oblique position would increase vulnerability. (*See* Plates XXII, XXIII, and XXVI.)*

Section 50.—Firing in other positions.

256. The recruit will also be instructed to fire:—

i. Sitting, which is most suitable when on ground falling at a steep slope. In this position the right shoulder should be kept well back, and the left forearm supported by the thigh, the right elbow resting against the right knee, or unsupported, as desired. (*See* Plates XXXI, XXXII, and XXXIII.)*

ii. Kneeling, used mainly when firing from continuous cover, such as a low wall, bank, or hedge, or in long grass, crops, &c., which would obstruct the line of sight if the prone position were adopted.

The soldier may kneel on either or on both knees. In the former case the body may be supported on the heel or not, as desired; the left knee will be in advance of the left heel, and the left elbow rest on or over the left knee; the left leg, hand, and arm, and the right shoulder, should be in the same

* Following page 104.

vertical plane when firing in the open kneeling on the right knee. In the latter, the body may rest on both heels, or be kept upright to suit the height of the cover, the elbows in both instances being unsupported by the body. (*See* Plates XXIV, XXV, and XXVII to XXX.)*

Section 51.—*Use of cover.*

257. Besides teaching the firing positions which are described above, the instructor will explain to his squad the best means of using various forms of cover for fire effect and protection. Cover for use during firing exercises will always be provided.

258. In the intervals of firing in the open the head should be lowered, but ground in front must still be watched by selected observers.

259. The value of cover from view and the means of concealment afforded by small folds in the ground, a few tufts of grass, &c., will be illustrated, and the tendency to attract attention by exaggerated movements of the head, arms, or rifle, in loading and aiming, will be pointed out and checked.

Section 52.—*Rapid loading.*

260. When he is able to aim and fire steadily in all positions and from various classes of cover, the soldier will be exercised in combining rapid loading with the greatest rapidity of aim consistent with accuracy.

261. Rapid loading will first be practised separately, using dummy cartridges in chargers. When five rounds have been

* Following page 104.

inserted in the magazine, the bolt will be closed and turned over, and the rifle will be at once unloaded and another charger inserted similarly. Rapid loading should be practised in all positions, but especially in the prone position.

262. Before dummy cartridges are used on parade, special precautions will be taken to ensure that neither ball nor blank ammunition is taken to the parade ground. The instructor will personally examine all cartridges, rifles, pouches, and bandoliers, before loading takes place.

263. The rate of firing should be increased gradually, provided that faults of aiming and trigger pressing are not acquired. Short bursts of rapid fire only will be permitted, the firing being carefully regulated and controlled. A target will always be named.

Section 53.—Use of the safety catch and cut-off.

264. Troops armed with rifles fitted with safety catches will invariably set the catch to safety before movement. The use of the cut-off is to be confined in their case to occasions when they are not actually engaged with the enemy, when it may be employed for the purpose either of charging the magazine without inserting a cartridge in the chamber, or to unload the rifle while retaining cartridges in the magazine. It is never to be used to enable the rifle to be used as a single loader, and is not to supersede the use of the safety catch.

265. In the case of rifles which have no safety catches, the cut-off will be pressed in and the rifle unloaded on all occasions

Plate XIX.

Standing.

Position when Loading.

Points to note:—

1. Body erect and well-balanced.
2. Left elbow close to body.
3. Firm grip with left hand, close in front of magazine.
4. Muzzle pointing upwards.
5. Butt well forward.
6. Forefinger of the right hand under the cut-off.
7. Eyes on the mark.

Plate XX.

Standing

Position when Firing.

Points to note:—

1. Body well-balanced.
2. Left elbow well under rifle.
3. Good bed for the butt.
4. Firm grip with both hands.
5. Eye well back from the cocking-piece,
6. Sights perfectly upright.

PLATE XXI.

STANDING.

Position when Firing

Points to note:—

1. Body well-balanced
2. Left elbow well under rifle.
3. Good bed for the butt.
4. Firm grip with both hands.
5. Eye well back from the cocking-piece.
6. Sights perfectly upright.

Plate XXII.

Lying.

Position when Firing.

Points to note :—

1. Body oblique to line of fire.
2. Legs separated.
3. Heels on ground.
4. Good bed for the butt.
5. Firm grip with both hands.
6. Eye well back from the cocking-piece.
7. Sights perfectly upright.

PLATE XXIII.

LYING.

Position when Firing.

Points to note:—

1. Body oblique to line of fire.
2. Legs separated.
3. Heels on ground.
4. Good bed for the butt.
5. Firm grip with both hands.
6. Eye well back from the cocking-piece.
7. Sights perfectly upright.
8. Elbows closed slightly inwards.

Plate XXIV.

Kneeling.

Position when Firing.

Points to note:—

1. Body well-balanced.
2. Left elbow well under rifle.
3. Good bed for the butt.
4. Firm grip with both hands.
5. Eye well back from the cocking-piece.
6. Sights perfectly upright.
7. Left heel slightly behind left knee.

Plate XXV.

Kneeling.
Position when Firing.

Points to note:—

1. Body well-balanced.
2. Left elbow well under rifle.
3. Good bed for the butt.
4. Firm grip with both hands.
5. Eye well back from the cocking-piece.
6. Sights perfectly upright.
7. Left knee, forearm, rifle, and right shoulder, in one vertical plane

PLATE XXVI

LYING BEHIND ISOLATED COVER.
Position when Firing.

Points to note:—

1. Body and legs parallel to the line of fire.
2. Side of rifle rested.
3. Backsight and dial sight not touching cover.

Plate XXVII.

Kneeling behind Cover.
Position when Loading.

Points to note:—

1. No undue exposure.
2. Watching front.
3. Trigger guard downwards to facilitate loading and sight setting
4. Muzzle clear of cover.

PLATE XXVIII.

KNEELING BEHIND COVER.
Position when Loaded.

Points to note:—

1. No undue exposure.
2. Watching front.
3. Trigger guard outwards.

PLATE XXIX.

KNEELING BEHIND COVER.

Position when Firing.

Points to note:—

1. No undue exposure.
2. Body well-balanced.
3. Left elbow well under rifle.
4. Good bed for the butt.
5. Firm grip with both hands.
6. Eye well back from the cocking-piece.
7. Sights perfectly upright.
8. Rifle, but not hand, resting on cover.
9. Position adapted to the cover.

PLATE XXX.

KNEELING BEHIND COVER.
Position when Firing.

Points to note:—

1. No undue exposure.

PLATE XXXI.

SITTING.
Position when Loaded.

PLATE XXXII.

SITTING.
Position when Firing. Aiming across a valley.

PLATE XXXIII.

SITTING.

Position when Firing. Aiming down a steep slope.

when the safety catch is ordered to be applied in these instructions. In an advance in extended order, however, these rifles may be carried during movement at the "slope" instead of being unloaded.

Section 54.—Muscle exercises.

266. To accustom the muscles to the strain of prolonged firing the following exercises will be performed daily during the elementary training of recruits and frequently by trained soldiers. Care must be taken that men are not unduly fatigued.

In each practice, a conspicuous object, representing the target, will be indicated, and the rifle will invariably be thrown into approximate alignment with it. In the first and third exercises the correct aiming position will be assumed, including taking the first pull, bringing the cheek on to the butt, and closing the left eye, but without actually aligning the sights. In the second exercise the first pull will be taken when the right hand grasps the rifle, but the head will not be lowered, the left eye will not be closed, nor will the sights be aligned.

i. 1*st Practice.*

(To be performed with and without bayonet fixed.)

Caution :—*Muscle Exercise.* 1*st Practice.*

Standing. Load.
(*Lying.*)
One.—Bring the rifle to the position for aiming, return at once to the position for loading, and continue the practice.
Unload.—As before.

ii.—*2nd Practice.*

Caution :—*Muscle Exercise. 2nd Practice.*

Standing. Load.
(*Lying.*)
One.—Bring the rifle to the position for aiming.
Two.—Quit the rifle with the right hand.
Three.—Seize the rifle with the right hand, and at the same time quit it with the left hand.

Note.—The words *Two* and *Three* will be given at intervals of about 10 seconds. The trigger will not be pressed when in the third position.

Unload.—As before.

iii.—*3rd Practice.*

(To be performed with and without bayonet fixed.)

Caution :—*Muscle Exercise. 3rd Practice.*

Standing. Load.
(*Lying.*)
One.—Bring the rifle to the position for aiming.
Unload.—As before.

Note.—The men will be trained progressively to hold the rifle in this position until they can do so without fatigue for two minutes.

Section 55.—Fire discipline.

267. When the recruit has profited sufficiently by the lessons of aiming and firing instruction, he will be practised in moving in extended order in accordance with the instructions contained in the training manual of his arm.

268. Fire opened without orders from a fire leader is called individual fire; only exceptional targets and very favourable atmospheric conditions will justify an individual skirmisher in opening fire at distances beyond (about) 600 yards.

269. While fire discipline should be such as to produce satisfactory results from individual fire, it should also ensure implicit obedience to orders for fire direction, collective action being necessary to give reasonable assurance of fire effect when opening fire at distances beyond (about) 600 yards, as well as to form a cone of fire for observation and correction of sighting.

Section 56.—Collective fire.

270. When the fire of several rifles is combined for a definite purpose under the orders of a fire leader, it is called collective fire; the senior officer or non-commissioned officer who defines the purpose for which fire is to be used is said to direct the fire, and it is the duty of subordinate commanders to control the fire in accordance with his orders.

271. Collective fire may be concentrated or distributed. Concentrated fire produces a cone of fire favourable to observation of results, and is more effective than distributed fire at the point of application. Fire may be distributed laterally (sweeping) or in depth (searching). Fire distributed laterally is to be preferred for neutralising an enemy's fire along any portion of his front, and fire distributed in depth gives greater assurance that some portion of the fire will be effective when the target has not been definitely located, or serious errors in sighting are to be expected. Fire used to cover movement or entrenchment should be carefully and systematically distributed. Against exceptionally vulnerable targets fire should be concentrated in order that its effect may be more clearly observed.

272. Collective fire is necessarily controlled fire; individual fire will also be controlled if circumstances permit.

Section 57.—Rates of fire.

273. The rate of fire will always be carefully regulated according to tactical requirements. Slow, desultory fire may disturb the enemy's aim, but it is opposed to principles of surprise. Short bursts of rapid fire may surprise the enemy before he can take cover, they favour observation of results, and afford intervals of time for adjustment of sights and fire discipline. The duration of such bursts must be strictly controlled. (*See* "Infantry Training.") In order to ensure control and to facilitate the passing of orders, the number of rounds to be fired may be named, *e.g.*, "5 rounds fire or rapid fire."

274. Soldiers working in pairs for observation and mutual assistance may *each* fire about three rounds a minute. The rate of deliberate fire should not exceed six rounds a minute. In rapid collective fire the rate will vary according to the visibility of the aiming mark, the range and the standard of training a man has reached. With a distinct aiming mark within about 1,000 yards a well-trained man should be able to fire from 12 to 15 rounds per minute without serious loss of accuracy.

275. If rapid fire is ordered, every man will fire at his own best rate for combining rapidity with accuracy.

276. Preliminary collective exercises will consist of simple practices designed to inculcate implicit obedience and prompt action on the instructor's command, and to practise rapid concentration of fire on different objectives.

Section 58.—Description of targets.

277. Collective fire cannot be effective unless the objective is described in such a way that every individual in the fire unit will immediately recognize his target or point of aim.

278. As targets will often be a fold in the ground or patch of open ground without any definite distinguishing mark, some system of describing targets is a necessary part of fire direction, and there should be only one system in use in any battalion.

279. At the conclusion of a movement, or when occupying a position, officers commanding companies will select one or more prominent objects as may be necessary in the front as description points approximately two hands apart, and impress upon all the names by which these objects will be known in directing fire. When possible, no two of these points should be of a similar nature. Targets will, when necessary, be described with reference to these description points, lateral distance from the description points being measured by the breadth of one or more fingers held vertically at arm's length in front of the face. When describing a target immediately above or below a description point, the fingers will be held horizontally, but for targets bearing obliquely from a description point the fingers will be held vertically and the bearing notified in degrees, or more simply, by reference to the position of figures on the dial of a clock, the description point being regarded as the centre of the dial.

280. Landscape targets should be used for preliminary lessons in the application of collective fire. Instruction may be given by means of these targets in the description and recognition of features of ground, and common objects in a landscape. The use of field glasses should be practised, and the targets will be found useful for explaining methods of attack and defence, the use of hedgerows as covered approaches, or methods of crossing open ground. Fire positions, sites for intrenchments, dead ground, positions for machine guns, areas likely to restrict fire owing to obstruction in the line of sight, ground to be crossed at top speed, &c., may be indicated as a preliminary to outdoor work.*

Section 59.—*Words of command.*

281. Words of command will be as few as possible. When practicable, directions as to target and sighting will be given before the occasion for firing arises, and fire will be opened without further orders as soon as the target appears. It will often be best to give orders for adjusting the sights first, so that there may be no necessity for the firers to remove their eyes from the target after it is named to them, but the order of the words of command is not of material importance. It is necessary to anticipate events, so that lengthy orders will not be needed after the target appears.

282. If all ranks are kept informed of the course of events and led to anticipate occasions for fire action, there should be no need for any words of commands other than those which regulate movement, the opening and closing of fire, and the rate of fire, and even these may be dispensed with if the firers are well trained and combine their efforts according to orders issued in anticipation.

283. The results of observation and alterations to be made in sighting or point of aim must, however, be notified at once. If firing tends to become wild, it should be stopped, and

* For further details as to landscape targets *see* Part II of these Regulations.

resumed only under strict control and detailed orders. Every man will in any case press his trigger independently, volleys being less effective than bursts of independent fire.

284. The following words of command will be used as may be found necessary :—

At—(Object).

At—(Elevation and deflection).

Fire or Rapid Fire	On which the firer will load, adjust his sights, aim and fire, slowly or rapidly.
Cease Fire	On which fire will be discontinued, and the firer will bring the rifle to the loading position, re-charge the magazine, and apply the safety catch.
Rest	On which the safety catch will be applied and an easy position assumed. When rifles are not provided with safety catches, on the command "*Rest*" the cut-off will be pressed in, the bolt opened and closed, and the spring eased.
Unload	On which all cartridges will be removed from the chamber and magazine, and other motions performed as detailed in para. **243**.

*Section 60.—Preliminary exercises.**

285. The squad will be placed in line at one or two paces interval, and on the command from the instructor ("*Standing*"

* Instructions for advanced exercises are contained in "Infantry Training."

"*Sitting,*" "*Kneeling*"); *At* (object); *At* (range); "*Fire*" or "*Rapid Fire,*" will perform the necessary motions, and continue firing until the command "*Cease Fire*" or "*Unload*" is given. If no orders are given as to the position, the squad will lie down.

286. The standing, kneeling, and sitting positions will only be practised in conditions suitable to their employment.

287. If it is desired to change front or position, or to vary the objective, range, or rate of fire, the instructor will give the necessary commands, but without, as a rule, causing fire to cease.

288. All commands given during firing are to be passed down the line of firers.

289. The safety catch will be applied or rifles will be unloaded (or sloped) before a movement is undertaken.

290. When the squad has gained sufficient experience it will be taught, when in motion, to halt, take cover if possible, and deliver fire on the executive word "*Fire*" or "*Rapid Fire,*" preceded by the necessary cautions.

291. To develop individuality, the complete detail of commands will occasionally be dispensed with. Thus at an object appearing suddenly for a limited time, the executive command "*Fire*" or "*Rapid Fire*" only will be given, on which each individual will adopt the position he considers most suitable to the tactical conditions, adjust his sight, and open fire. The instructor will observe and criticise the positions and the sighting of the rifles.

292. Anticipatory orders will sometimes be given; for example—'the enemy is about to advance out of the firwood; when he moves, concentrate on the thickest part of his line, 1550 and 1650'

Section 61.—*Fire in two ranks.*

(For use in savage warfare.)

293. Cautionary commands by the company commander—

(BY PLATOONS) (BY SECTIONS) AT (OBJECT), AT (RANGE). FIRE (OR RAPID FIRE).

Fire control orders by platoon or section leaders—

i. *Fire* (*or Rapid Fire*).

The rear rank closes on the front rank, taking a pace to the front with the left foot and a short pace to the right front with the right foot. The front rank kneels, the rear rank stands.

ii. *Unload.*—After ordering arms the rear rank resumes its original position.

294. The normal firing positions in two ranks are rear rank standing, front rank kneeling. If it is desired to fire both ranks standing, or both ranks kneeling, or front rank lying, rear rank kneeling, the directions to do so will precede the command *At—(object), &c.*

Section 62.—*Webley pistol exercise.*

295. All officers, warrant and non-commissioned officers and men armed with revolvers in war are to be frequently instructed in the mechanism and practised in the use of the revolver.

i. *Single Practice.*—The instructor, having previously taught the names and parts of the pistol (Sections 7 and 8), will form up the squad at one or two paces interval, about 15 or 20 paces from the marks to be aimed at—

Load

Draw the pistol with the right hand, pointing the muzzle towards the target. Seize the barrel with the left hand, knuckles upwards, press forward the lever with the thumb of the right hand, and open the pistol with a jerk.

Hold the pistol in the left hand, and insert a cartridge in each chamber. Close and rest it on the palm of the left hand, muzzle pointing towards the target, the forefinger extended on the body.

—— rounds with right (or left) hand. Single practice

Right hand practice.—Make a half-left turn, keeping the muzzle pointed as before. Advance the right foot about 10 inches, and full cock with the thumb of the right hand.

Left hand practice.—Make a half-right turn. Advance the left foot about 10 inches; reverse the pistol, holding it in the left hand and resting it on the palm of the right hand, and full cock with the thumb of the left hand.

Fire

Dropping the left (or right) hand to the side, raise the pistol, at the same time placing the forefinger round the trigger; take aim through the notch on the breech, aligning the foresight on the mark. As soon as a correct aim is obtained, press the trigger until the hammer falls, return to the loading position, full cock, and continue the practice until the named number of rounds is expended. Return to the loading position, and, if at left hand practice, reverse the pistol to the palm of the left hand.

When aiming and firing the butt should be firmly held.

Unload	Open the pistol as before detailed when the empty cases will fall out (unfired cartridges will be removed by opening the pistol partly and only with sufficient force to raise them from the cylinder). Close the pistol.
Return—Pistol	Turn to the front, and return the pistol to its case.

ii. *Continuous practice.*

Load.	As before detailed.
—— rounds with right (or left) hand. Continuous practice	As before detailed, but without cocking.
Fire	Dropping the right (or left) hand to the side, raise the pistol, at the same time placing the forefinger round the trigger, take aim through the notch on the breech, aligning the foresight on the mark ; as soon as a correct aim is obtained, press back the trigger to its full extent ; after firing release the trigger with the forefinger, and, without full cocking, continue to fire until the whole of the chambers are discharged ; then return to the loading position. After firing with left hand, reverse pistol to right hand.
Unload.	As before detailed.
Return—Pistol.	As before detailed.

iii. The command *Cease Fire* will be used only when it is necessary to cease fire before the named number of rounds is expended.

Section 63.—Tests of elementary training.

296. Range practices and more advanced training may be regarded as waste of ammunition and time unless recruits have been thoroughly grounded, and trained men are kept efficient, in elementary training. The following system of tests has therefore been designed in order to :—

i. Provide instructors with a means of testing recruits to ensure that they have reached a sufficient standard before they begin range practices.

ii. Ensure that trained soldiers have retained their efficiency.

iii. Prevent any detail of elementary training from being overlooked.

iv. Provide a standard to be attained by technical and other troops who are unable to devote as much time as is desirable to elementary training.

v. Enable non-commissioned officers and men, particularly those of the Special Reserve and Territorial Force, to reach the required standard in their own time.

These tests are divided into oral, inspection and standard tests. A record will be kept for each man of the results of the various tests, which will be inspected periodically by the commanding officer. Extracts from these records will furnish useful guides as to efficiency when men are transferred to other companies or battalions. Men, particularly recruits, should themselves keep a record of their performances.

It is important that teaching should not be confused with testing. In the former a man is instructed by example and explanation; in the latter he is questioned, or ordered to carry out a certain exercise without any explanation or assistance, and either passes the qualifying standard or is relegated for further instruction.

Section 64.—*Method of conducting the tests.*

297. *Oral tests.*

i. *Care of arms and ammunition.*

A few questions should be put to each man.

ii. *Description of natural objects.*

Each man separately should be called upon to describe one or two objects in the landscape, and be questioned as to shape, colours, sizes, units of measure, &c.

iii. *General theoretical knowledge.*

A few questions should be put to each man regarding the objects of, and reasons for, various details that have been taught him.

298. *Inspection tests.*

i. *Firing positions.*

Every man should be individually inspected in all firing positions and the existence of any of the following serious faults should be noted in a book for production at subsequent tests, viz., firing from left shoulder, eye near cocking-piece or thumb in aiming, want of grip with either hand, excessive constraint, finger round trigger in loading position,

ii. *Fire discipline.*

The rapid execution of all orders for fire direction, adjustment of sights after each advance in attack, and before every advance of the enemy (represented by fatiguemen) in decisive phases of engagement, should be tested.

299. *Standard tests.*

i. *Eyesight.*

To be carried out in conjunction with Standard Test iii. Four fatiguemen as "points" should be placed under cover in various directions and at different distances, not exceeding 800 yards. The men to be tested lie down extended to two paces. Each point is called up by signal.

The fatigueman stands, kneels, or raises his head, according to the degree of visibility required, and fires four rounds of blank ammunition in half a minute, then returning to cover. During the half minute the observers adjust their sights and place their rifles at arm's length to the front. At the end of the half minute a whistle is blown, and those men who have failed to discern the point are noted, while non-commissioned officers record the elevations found on the sights in connection with Standard Test iii. In every case the point should be signalled to rise a second time in order that his position may be shown to those men who previously failed to see him, and half a minute should then be allowed to these men to adjust their sights for Standard Test iii.

Failures not to exceed one per man, but consideration must be given to the visibility of the points.

ii. *Recognition of targets and aiming points.*

The men to be tested should each have an aiming rest or sandbags. A non-commissioned officer from behind them will describe some difficult aiming point, such as a point in a hedge or area of open ground. The men lay their rifles on the point which they recognise from the description. Four points should be described for every man tested.

iii. *Judging distance.*

The distances of four standing fatiguemen should be judged at distances not exceeding 800 yards.

iv. *Adjustment of sights.*

Several distances will be named and sights examined after 3 seconds [M.L.E. rifle, 5 seconds]; an extra 5 seconds will be allowed if deflection is ordered.

v. *Regulation aim.*

To be tested by triangle of error. Standard—No side over one-third of an inch, or the centre of the triangle more than one-third of an inch from the instructor's aim.

vi. *Trigger pressing.*

Trigger pressing will be tested by means of the aim corrector, or sub-target machine. The Standard, if the sub-target machine is used, will be :—Marksmen, ·2 inch group; 1st Class, ·4 inch group; 2nd Class, ·6 inch group.

vii. *Aiming off for wind or movement.*

Tested for wind with the aiming rest. The men will be ordered to lay their rifles on a point at some number of feet, not exceeding 6, right or left of a fatigueman. One foot of error only, measured from the regulation point of aim, will be allowed.

Allowance in aiming off for wind or movement will be tested with the aim corrector, the percentage of serious errors being recorded.

viii. *Rapidity of aim.*

The time required to bring the rifle from the loading position to the shoulder, on the command "*Fire*," and to align the sights on an aiming disc held to the eye, will be measured with a stop or ordinary watch with a second hand. Position, prone.

The instructor will stop the watch when the trigger is pressed, provided he is satisfied with the aim. Four seconds will be the standard time. The subtarget rifle may be used for this test.

ix. *Rapidity of loading.*

The men to be tested will be equipped with a bandolier, pouch or cartridge pockets, and six chargers filled with dummy cartridges.

The chargers will be placed in the pouches or pockets, which will be buttoned over them.

The time required to load, close the bolt, eject the cartridges, the rifle being held in the correct loading position, one charger being inserted at a time, the pouch or pocket, whether empty or not, being buttoned up every time a charger is withdrawn, will be noted. Standard time, 1 minute.

x. *Rapidity of firing.*

This will be a combination of viii. and ix. On the command "Rapid fire," each man will load with dummy cartridges in chargers from the pouch or cartridge pocket, the pocket being buttoned up each time a charger is withdrawn, and aim 10 rounds at an aiming disc held to the eye. If the aiming is unsatisfactory, the test will be repeated more slowly. The time required will be noted. Standard time, 1 minute.

xi. *Grouping with miniature cartridges.*

For Regular troops, the grouping standards for miniature cartridge practice at 25 yards will be:—Marksman, 1-inch ring; 1st Class, 2-inch ring; 2nd Class, 3-inch ring; and for the Special Reserve and the Territorial Force:—Marksman, 2-inch ring; 1st Class, 3-inch ring; 2nd Class, 4-inch ring.

300. *Cancelled.*

CHAPTER V.

VISUAL TRAINING AND RANGING.

Section 65.—*General.*

301. The principal means of ranging are i. judging distance by eye; ii. observation of fire; iii. range finding; but there are several auxiliary methods such as use of maps, cross bearings, back-reckoning, information obtained from artillery or machine guns, sound and flash, &c., which may be used exceptionally. No means of ascertaining ranges should ever be neglected, if time and opportunity are favourable. Range finding will, as a rule, be in progress before fire is opened, and the rapid reconnaissance of any ground on which a fire fight is likely to take place should provide information of the greatest value to company officers responsible for the direction of fire.

302. Errors in sighting cause a greater loss of fire effect than personal errors in shooting, except at close range, but even within close range mistakes of more than 100 yards are not infrequently made in judging distance. The more accurate the firing the less will be the result if the sighting is incorrect.

303. As fire is controlled at longer ranges, practice in judging distance is more necessary for officers and non-commissioned officers than for the private soldier, though any soldiers who show aptitude in judging short distances should be given training with their leaders at greater distances.

304. The mean error of private soldiers in judging distances within 800 yards range should not exceed 100 yards.

305. Officers and non-commissioned officers, by constant practice, will reduce their mean error in judging distance from about 20 per cent. of the correct distance to about 10 per cent., but much depends on the local conditions to which the observer is accustomed, and serious errors must be expected in judging under strange conditions of ground and atmosphere. Exercises in judging distances must therefore be carried out during as well as before a campaign in a country overseas.

306. Great as is the loss of fire effect due to errors in judging distance, an even greater amount of ammunition will be expended without good result unless the firer is trained to discern and discriminate between targets at close ranges, and to recognise targets described to him at effective and longer ranges.

Section 60.—*Visual training.*

307. Exercises framed to stimulate the soldier's powers of discernment and recognition should be commenced at the earliest period of his training, and continued throughout his service, in order that he may appreciate, by daily association, the importance of cultivating a quality of such importance in modern warfare.

308. Visual training will include general reconnaissance, discernment of targets, aiming, impressions of size, recognition of targets described by word of mouth, and observation of the results of fire. Training will begin with questions framed to develop the recruit's powers of discerning objects and describing what he sees. Any ordinary objects will be counted, and figures of different colours will be placed sometimes in the open and sometimes under partial cover in front of various backgrounds. Fatiguemen will be employed to skirmish from cover to cover and to perform the firing motions from behind cover, in order to show how motion catches the eye and exposes the firer's position. Blank ammunition will be used to give practice to the ear in locating an enemy by sound.

309. As progress is made, these exercises will be carried out under stricter conditions, the observer lying down, and the figures being of neutral tints. Special attention will be given to recognising features of ground such as fire positions, dead ground, areas restrictive of fire effect, &c. The use of field-glasses will be practised, objects in the distance being examined and described, both with and without the aid of glasses.

310. In connection with the visual training recruits will be familiarized with all terms applied to features of ground, colours, shapes, measurements, and military objects generally, so that their powers of description may be improved for reconnaissance purposes, and that their faculty of recognizing targets from description may be developed for purposes of collective fire.

311. Visual training should at first be separated from training in judging distance, but as proficiency is attained both should be combined as a preliminary to the fire discipline and collective fire exercises already referred to. (Sections 55 and 56.)

Section 67.—*Judging distance.*

312. Distances may be judged i. by measuring the intervening ground with the eye in terms of some familiar unit such as 100 yards; ii. by the apparent size of the object if its size is known (visual angle); iii. by the visibility of the object as affected by light, atmospheric effect, background, &c.

313. All methods should be practised until it is found that distances can be approximately judged from the general impression conveyed to the eye, but the observer must bear in mind that his judgment may be influenced by certain conditions of ground, light, &c., which are mentioned below.

314. The recruit will first be familiarised with short units of distance not exceeding 600 yards, and this distance (600 yards) he should particularly recognise as the limit of individual fire. He will at the same time study the visibility of the human figure standing, kneeling, and lying, at known distances. He will be required to remember the results of his observation, and be given further opportunities of seeing figures in varying conditions of light, atmosphere, and background.

315. He will afterwards be taught the influence exercised by ground, and features in the landscape, on general impressions of distance.

i. *Objects are over-estimated—*

When kneeling or lying.

When both background and object are of a similar colour.

On broken ground.

When looking over a valley or undulating ground.

In avenues, long streets, or ravines.
When the object lies in the shade.
When the object is viewed in mist or failing light.
When the object is only partially seen.
When heat is rising from the ground.

ii. *Objects are under-estimated*—
When the sun is behind the observer.
In bright light or clear atmosphere.
When both background and object are of different colours.
When the intervening ground is level or covered with snow.
When looking over water or a deep chasm.
When looking upwards or downwards.
When the object is large.

316. Practice will follow in judging the distances of natural objects, features of ground, and fatiguemen, reasons being given for his estimate by each observer, and the local conditions being carefully examined and explained by the instructor before passing on to the next object.

317. Assistance may be obtained by making a maximum and minimum estimate of the distance, and taking the mean for correct, trying to halve the distance and judge the half distance first, or by judging the distance of some object of known size which appears to be in the vicinity of the objective, should the latter be a fold in the ground or of unknown size.

318. As progress is made the time allowed for judging distances should be limited, but rough guessing is never to be allowed. Further training should include deduction as to distances drawn from any known ranges previously communicated, estimation of lateral distances, and measurement of the visual angle by means of a protractor, the fore-sight, the barrel of the rifle, or similar method.

319. Officers and non-commissioned officers should be given occasional opportunities of revising their impressions of the human figure at known distances, and should be encouraged to use small-scale maps as a guide in judging.

320. Field sketching and reconnaissance afford excellent training in judging distance.

321. Exercises in ranging by means of maps, range finders, and the eye, used in combination, should form an essential part of the fire direction practices of the annual course of musketry.

Section 68.—*Tests in judging distance.*

322. During tests in judging distance no assistance from maps or other means should be allowed.

323. In every company, and at depôts, in addition to instructional exercises, at least one test in judging the distances of four objects will be held for all ranks in every three months, the distances of the objects being between 200 and 800 yards.

324. All company officers, non-commissioned officers, and those men who are in possession of badges for judging distance, will be tested also once in every three months under regimental arrangements in judging the distances of four objects between 200 and 1,400 yards.

325. Brigade commanders and officers commanding districts will issue such orders as may be necessary for exempting non-commissioned officers and men from these tests, but every officer and serjeant who would command a company or smaller unit on mobilization, should attend four tests annually in judging distances up to 1,400 yards.

i. The tests will be conducted on unfamiliar ground, half the objects consisting of fatiguemen or dummy figures representing skirmishers, and half of natural objects marking fire positions, such as would be occupied on service.

ii. The observers will lie down or kneel behind cover, and estimates will be recorded by adjusting the sights (or in the case of officers by writing) in multiples of 50 yards; the objects should have been previously placed or selected, or the position of the observers should be chosen, so that the correct distance may closely approximate to some multiple of 50 yards.

iii. Half a minute will be allowed for each estimate, reckoned from the moment when the object is pointed out or a shot is fired to draw attention to the position of the object.

iv. At the conclusion of the half-minute a whistle should be blown, when the observers will stand to attention, and no further adjustment of the sights or writing will be permitted.

v. The register keeper will then examine each rifle or paper, and record the estimates in the register.

vi. On return to barracks, the mean percentage of error of each observer will be entered in the register, which will be preserved for record.

326. On the completion of the classification practices, the mean percentages of error made by each individual officer, non-commissioned officer and man since the last classification will be added together and divided by the number of tests which he has attended, the average thus obtained being regarded as the standard of proficiency of the individual until the next classification.

327. Officers, non-commissioned officers, and men whose mean error exceeds 20 per cent., and those who have attended less than two tests, will be regarded as inefficient.

328. Additional practice will be afforded weekly to all officers, non-commissioned officers, and men who are recorded as inefficient, until the next quarterly test.

329. Officers and serjeants whose mean error exceeds 20 per cent. will not be classified as better than 2nd class shots.

Section 69.—Range finding by observation of fire.

330. Observation of results is the best means of correcting errors in sighting, but it is only practicable when the ground in the vicinity of the objective is of a nature to show the strike of the bullets, or when the enemy is in the open.

331. A considerable volume of rapid and concentrated fire is necessary to enable the strike of bullets to be observed, and for this purpose one or more Maxim guns or not less than half a company of infantry, or its equivalent, should be employed. The distance, and nature of the ground, will determine the volume of fire required, but the greater the difficulty of observation the greater must be the volume of fire. When trying for observation, 3rd class shots should not be permitted to fire.

332. An elevation well under the estimated distance should first be selected, and if the fire can be observed, the elevation should be increased by not less than 100 yards at a time until the nucleus of the bullets is seen to fall on the desired spot.

333. The point at which the fire is directed may be the actual objective or ground in its vicinity which is more suited to observation of the strike of the bullets.

334. The best position from which to observe fire is behind, and, if possible, above the firers, but in this position the bullets which fall short will be most easily seen, and may be mistaken for the nucleus of the fire; further, all shots will appear to strike nearer to the observer than is really the case.

335. To an observer on or behind a flank of the firers, shots which pass over the mark will appear to fall towards the side on which he is posted, and those which drop short to fall towards the opposite side. Thus, if the majority of shots seem to an observer on the right flank to fall to the right of the mark, the range has been over-estimated, and if to the left, under-estimated.

Section 70.—*Test of range-finding instruments.*

336. All range-finders will be tested once in every three months at least under arrangements made by brigade commanders, or by officers commanding districts in the case of depôts. A fully qualified officer will be selected to superintend; he will verify ranges before the instruments are tested. Each instrument will be tested for accuracy and consistency in taking at least four ranges.

Section 71.—*Training and test of range-takers.*

A. *Regular Forces.*

337. In every regiment of cavalry and battalion of infantry, the following establishment of trained range-takers will be maintained. A due proportion of range-takers will also be maintained at cavalry and infantry depôts :—

ESTABLISHMENT OF TRAINED RANGE TAKERS.

Units equipped with one-man range-finder only.

Arm of Service.	Officers.	N.C.Os. or privates.	Remarks.
Cavalry	3*	8†	* One will be the machine-gun officer.
Infantry	3*	10†	† Two privates for the machine-gun personnel.

Units equipped with the mekometer only.

Arm of Service.	Officers.	N.C.Os. or privates.	Remarks.
Regular Forces—			* One of these will be the machine-gun officer.
Cavalry	3*	14†	† Two of these men will belong to the machine-gun personnel.
Infantry	3*	26†	
Special Reserve—			
Cavalry	3*	14†	
Infantry	3*	26†	
Territorial Force—			
Cavalry	3*	14†	
Infantry	3*	26†	

Units equipped with both instruments.

Arm of Service.	Officers.	N.C.Os. or privates.	Remarks.
Mekometer—			* These will be the same officers as are trained in the mekometer. One will be the machine-gun officer.
Cavalry	3	9	† Two of these men will belong to the machine-gun personnel.
Infantry	3	12	
One-man—			
Cavalry and infantry	3*	4†	

338 (*a.*) The three officers in each cavalry regiment and infantry battalion will be required to qualify at a range-taking course at a school of musketry.

(*b.*) The whole of the establishment of non-commissioned officers and men to be maintained as range-takers in a cavalry regiment and an infantry battalion will be trained regimentally.

(*c.*) No man above the rank of corporal will be trained as a range-taker.

(*d.*) Range-takers, when once trained, should only be changed for special reasons. The establishment of officers qualified at a school of musketry and non-commissioned officers and men trained regimentally should not be allowed to fall short of the numbers given above. Additional officers and non-commissioned officers and men will be trained in anticipation of vacancies.

(*e.*) An officer who has qualified at a rifle or machine-gun course at which instruction in the one-man range-finder was given, will not be allowed to count such qualification for the purpose of this section. He will be required to qualify at a range-taking course.

339. Brigade commanders will arrange for the personnel forming the regimental establishment of range-takers to be tested once in every six months to ensure that a sufficient standard is maintained. They will also ensure that the numbers of range-takers are kept up to establishment, and that they are thoroughly efficient.

Officers commanding districts will arrange for officers and non-commissioned officers at depôts to be tested every six months.

It is most important that tests should be conducted by fully qualified officers who have had sufficient experience to be able to recognize, for the purposes of a test, the difficulty of ranging on and writing descriptions of objects selected, so that the test, as regards time limit and other conditions, should neither be too difficult nor too easy.

In addition to the test mentioned below, officers and non-commissioned officers of the range-taking establishment should possess—(*a*) a thorough knowledge of the hand-book of the instrument; (*b*) ability to instruct—this can generally be ascertained by the standard reached by those who have been trained regimentally; (*c*) capacity for handling the instrument practically, special attention being paid to the selection of a suitable spot from which to range with regard to ground and troops in the vicinity; selection of suitable marks on which to range; position adopted which should afford greatest concealment combined with steadiness.

Tests with the mekometer range-finder will be carried out in accordance with para. **340.**

Tests with the one-man range-finder will be carried out as follows :—

(1) The ranges to be taken will be between 600 yards and 2,000 yards, one to be beyond 1,400 yards and one within 1,000 yards. The correct ranges to objects selected will be ascertained by the superintending officer by taking the mean of a series of readings by at least 3 trained range takers with one or more instruments, which have been previously checked on the spot both for halving and zero adjustments. Great care must be taken to ensure that the "correct" ranges are accurate.

(2) The superintending officer will satisfy himself that the instruments to be used for a test are correct as regards halving adjustment and are otherwise in order.

(3) The range-taker will himself check, and if necessary adjust the instrument he is about to use for zero. He will be required to explain three methods, in addition to the method used, by which the zero can be adjusted if necessary. Officers who are being tested will also be required to explain how the halving is adjusted. Should a range-taker be unable to check the zero and explain three other methods, or an officer be unable to explain the halving adjustment, he will fail and will not be tested in taking ranges.

(4) Having satisfied himself as to the adjustment of the instrument, the range-taker will focus the telescope and prepare the instrument for use. The prone position will be adopted when the instrument is used horizontally and the most convenient position when used vertically.

(5) The superintending officer will indicate by description or by means of a rifle laid from a rest four fire positions or practical service objects the ranges to which are required. One of these, which will be specified, will be taken with the instrument in the vertical position. The positions or objects selected should be such that they can be taken and described within the limits mentioned in sub-paras. (6) and (8) below.

(6) When the positions or objects have been identified and the range-taker is ready, the word "go" will

be given, and 10 minutes will be allowed in which to take and write down the ranges, as well as a concise description of the object to which each range refers, so that the object can be easily identified by any commander to whom the range is handed. At least three readings should be taken to each object before the reading to be handed in is decided upon.

(7) The reading as well as the "correct" range will be recorded in multiples of 10 yards only.

(8) Should the time limit be exceeded or the average percentage of errors for the four ranges exceed 2·5 or for any one range 4, the range-taker will fail, otherwise he will qualify.

Brigade commanders will mention in their reports on units (para. 620 ix) whether the standard in range taking is satisfactory or otherwise, and if the required number of officers, non-commissioned officers and privates is maintained.

B. *Special Reserve and Territorial Force.*

340. All officers and non-commissioned officers of Special Reserve and Territorial Force units attending rifle courses will receive instruction in the use of the mekometer as heretofore.

Regimental range-takers (numbers as in para. **337**) will be tested under the conditions of para. **339** as follows:—

i. The range-takers will work in pairs, each pair being judged according to the results produced in combination.

ii. The superintending officer will ascertain the ranges of the objects selected with the instrument to be used in the test.

iii. The pair of range-takers to be tested will lie down under cover, the instruments being in their cases and the cord wound round the reel.

iv. The superintending officer will indicate the object to the range-takers, taking the time from the conclusion of this indication.

v. The range-takers will then take the range, observing service conditions as regards position, cover, &c.

vi. They may take as many readings as they wish, No. 1 at the conclusion giving the final reading as decided upon to the superintending officer, who will cause the time taken, as well as the reading, to be entered in the register.

vii. Readings, as well as the correct range, will be recorded in multiples of 10 yards only.

viii. The final reading given by No. 1 will be recorded in all cases, whether marks are allotted or not, so that the mean percentage of error in readings may be calculated.

ix. Each pair will take four ranges, between 600 and 2,000 yards, one to be beyond 1,400 yards and one within 1,000 yards, and each individual of a pair will use the reading instrument for two of the four observations.

x. Twenty-five marks will be allotted for each range observed—20 for accuracy, three for time, two for observance of service conditions.

xi. Marks will be deducted according to the following scale for time taken in excess of two minutes at any range :—

Not exceeding 15 seconds ...	Deduct 1 mark.
More than 15, but less than 30 seconds	„ 2 marks.

xii. If the range recorded has an error exceeding 5 per cent. of the correct distance as taken by the superintending officer, or if time in excess of 2 minutes 30 seconds is taken, no marks at all will be allotted for that range.

xiii. Marks will be deducted for inaccuracy as follows :—

For errors not exceeding 2% of correct range	No deduction.
For errors exceeding 2% and not exceeding 3%	Deduct 5 marks.
For errors exceeding 3% and not exceeding 4%	Deduct 10 marks.
For errors exceeding 4% and not exceeding 5%	„ 15 „

Those who obtain less than 50 marks in any test will be noted as inefficient, and will receive further instruction until reported efficient.

341. The mean percentages of error in all range-taking tests will be published from time to time, as commanding officers may direct, in Regimental Orders.

CHAPTER VI.

PRELIMINARY TRAINING AND RANGE PRACTICES.

Preliminary Training.

Section 72.—Preliminary training of recruits.

343. Although it is only by practice in shooting that a high degree of proficiency can be attained in the use of the rifle, the ammunition allowances are necessarily limited, and are calculated on the assumption that firing will be preceded by a most thorough course of preliminary instruction.

344. Recruits, before they begin a course of firing must reach a satisfactory standard in aiming, and in holding their rifles steadily while pressing the trigger; otherwise range practice will result only in waste of ammunition.

345. Shooting from the left shoulder is not to be permitted, unless it is rendered necessary by defective eyesight.

346. In the cavalry and infantry instruction should be progressive and thorough in every branch of preliminary training; the period of time required for recruits may be as much as 3 months, and will not be less than 6 weeks.

347. In the case of troops which cannot adopt a system of preliminary training lasting for 6 weeks or more, it is necessary to standardize all branches of training so that individuals may practise in their own time, with a view to satisfying tests applied before they are permitted to fire. (*See* Sections 63 and 64.)

348. The tests applied in the Territorial Force should speedily ensure a satisfactory standard of proficiency in those branches of training which cannot in ordinary circumstances receive further attention after the outbreak of war. Such branches of training are those which depend on range accommodation, special appliances, and specialized instruction.

349. Preliminary musketry training comprises instruction in firing under all conditions, and the application of musketry in varying tactical situations in accordance with the principles laid down in the Training Manuals. It should be supplemented by lectures on theory and tactics, and such drill or gymnastic exercises as may be necessary to develop will-power, and to co-ordinate the actions of the muscles under control of the brain.

350. The 30 yards or miniature range should be used to illustrate lessons of preliminary training as far as possible.

Section 73.—Preliminary training of trained soldiers.

351. Preliminary training in musketry exercises should be continued throughout the year by trained soldiers. Skill in judging distance, a perfect trigger release, dexterity in the loading motions and the habit of sight adjustment cannot be retained without frequent practise. It is of the utmost importance that trained soldiers should develop by constant use a habit of recognizing targets, judging their distance, adjusting their sights, and firing quickly but steadily without undue effort.

352. It is advisable to test the ability of trained soldiers in the standard tests of preliminary training before they begin firing in range practices. (Section 64.)

353. Reconnaissance and ranging exercises will be a necessary preliminary to successful fire direction.

Section 74.—*Miniature cartridge practice.*

354. Practice with miniature cartridges is a valuable preliminary to firing with service ammunition, provided that the rifles or aiming tubes possess a high degree of accuracy, and the methods of instruction are such as to expose and correct the faults of the firer.

355. During preliminary drill, frequent visits should be made to the miniature range, and the lessons of aiming, pressing the trigger, declaring the point of aim on discharge, &c., should be practically illustrated by means of miniature ammunition.

356. The principles followed should be those of practice with service ammunition; at first there should be exercises in grouping only until a 3-inch standard is reached. Methods of correcting error according to the result of the last shot fired should be illustrated in the early stages by using an elementary target; correction should afterwards depend on observation of fire aimed at a figure target set up on a sawdust bank.

357. The rifles used should be service pattern, ·22 inch R.F., or Aiming or Morris Tubes used in service rifles with regulation sights.

358. If the range is in a town, or if it is to be used by Territorial soldiers, there should be an equipment of landscape and figure targets. With the aid of these targets, instruction may be given in the description of ill-defined service objectives, such as areas of ground, also in methods of judging distance, justification for opening fire and collective fire theory.

359. The windgauge may be used to represent wind, and the firers taught to aim off so as to correct the deflection given, acting sometimes on their own judgment, sometimes according to orders for fire direction.

360. Magazines cannot be used unless adapters are available, and these are seldom satisfactory.*

361. Due regard should be paid to the visibility of service targets, and bull's-eye targets should be used for the first few rounds only.

362. It is desirable that firing should take place in the open air if possible, so that artificial conditions of light and visibility may be avoided.

363. Miniature ranges are well adapted for snapshooting practice at vanishing and moving targets.†

364. *Instructions for Employment of Aiming Tubes.*—i. No tube which does not bear the authorized Government view marks will be used in any Government rifle or carbine.

ii. Tubes may be employed by men who have been instructed in their use, under such supervision as commanding officers

---

* The War Office Miniature rifle used by some Cadet clubs is fitted with a magazine for miniature ammunition (*see* para. **52**).

† For further information as to Miniature ranges *see* Part II of these Regulations.

may consider sufficient to ensure that the arms will not suffer. They will be issued as required under regimental arrangements.

iii. As no injury to an arm can be effected by the proper use of tubes bearing the Government mark, any damage which may arise will be treated as a case of neglect, and charged accordingly. Bolts which have been adapted to take the rim-fire striker are not to be drawn back so that the bolt-head strikes the resisting shoulder. If they are brought forcibly into contact with the resisting shoulder, the bolt-head may be broken.

iv. Arms marked M.T. (Morris Tube), or A.T. (Aiming Tube), if on charge, are to be used with the tubes. The tubes and arms set apart for use on any miniature range will be in the care of the non-commissioned officer in charge.

v. Tubes which are private property will be examined at the Royal Small Arms Factory, Enfield Lock, upon application to the Chief Inspector of Small Arms, and will, if found correct in dimensions, &c., receive the proper view marks. A charge of one shilling per tube examined will be made, and the conveyance of these tubes to and from the factory will not be accepted as a charge against the public.

365. A non-commissioned officer will be placed in charge of each range, and will attend whenever any practice takes place.

Shooting will take place only during the hours fixed by the commanding officer.

366. The tube will be cleaned from the breech end after every 20 rounds fired.

367. No person, except the officer or non-commissioned

officer in charge, or the marker, is to pass from the firing point up to the target during practice. Should it be necessary to stop firing, the same precautions are to be taken as at rifle practice.

If precautions are necessary for the safety of the public while practice is proceeding on these ranges, the instructions given in Section 91 will be observed. When practice takes place on a classification range, the same orders for safety, &c., are to be observed as when service ammunition is used.

368. Extra aiming tube ammunition for voluntary practice may be drawn on payment, as laid down in the Equipment Regulations, Part I, 1909.

Section 75.—*Thirty yards ranges.*

369. When no classification range is available, elementary practices with service ammunition may be carried out on a 30-yards range. Such practice will render recruits familiar with the discharge of the rifle and improve their trigger release under easy conditions. No practice will take place unless an officer or experienced serjeant is present. All precautions for safety will be taken. Loading, in all positions except lying, will be carried out with the rifle held just above the waist and the muzzle directed towards the target. Charging or uncharging magazines is not to be carried out with the muzzle pointing upwards.

370. Practice at vanishing, moving, and landscape targets can be carried out as on miniature ranges, but with service ammunition.

Long range sighting targets should be provided as a means of ascertaining the error of the rifle, and practice may be carried out with long range sights.

371. 30-yards ranges are well adapted for the elementary practices of machine-gun detachments.

INSTRUCTIONAL RANGE PRACTICES.

Section 76.—*General.*

372. When recruits have shown clearly that they have acquired a satisfactory standard of skill in all branches of preliminary training, it is necessary that they should verify for themselves by practi e the lessons which, up to that time, have been taught in theory only.

373. The range practices for the Regular Army and Special Reserve begin with qualifying practices, for which standards are prescribed. If these standards are not attained, it is a sign that the preliminary training has failed in its object. After passing standard tests in sub-target grouping, miniature range practice, and firing on the 30-yards range, every man should begin his practice on the open range confident in his own powers, and determined to prove his ability to hit.

374. Instructional range practices need not be continuous; intervals of time between the exercises are often beneficial, especially to nervous men, but in any case time should be found to continue the exercises of preliminary training on days allotted to range practices, in order that there may be no separation between theory and practice.

375. In preliminary training the recruit has not accustomed himself to the shock of discharge. In some cases there is extreme difficulty in overcoming the tendency to flinch from this shock, and this is one of the commonest causes of inaccurate shooting. Men who flinch should not proceed with firing practice, the cause of the flinching can be detected and sometimes be removed after one or two rounds only.

376. It is a common experience that serious faults become formed habits in recruits before they are discovered by the instructor; and that they are exposed only after repeated visits to the range, by which time it will be difficult to correct them.

377. In order to compel analysis of faults, to indicate clearly how defects may be remedied, and to remove all suspicion as to the accuracy of his rifle, preliminary and qualifying practices are divided into exercises in (A) grouping, and (B) applying fire.

Section 77.—Grouping.

378. Individual grouping is an exercise in firing a series of shots (usually five) at a definite aiming point without alteration of aim or sighting during the series. As it is necessary to exclude atmospheric influences as much as possible, the range should not exceed one hundred yards.

The term "group" is used to define the diagram made on the target by the series of shot marks. The value of such a group is determined by the relative closeness of the shot marks. It is measured by means of rings of various sizes, and points are allotted according to the size of the ring which will contain all the shots, or, when specially prescribed, all the shots but one, which is called a "wide shot." When the ring is applied so as to contain all the shots, the point where the centre of the ring falls on the target is regarded as the point of mean impact. The position of the point of mean impact, with reference to the mark aimed at, has no influence on the value of the group. Its position is, however, of great importance for instructional purposes, because it indicates approximately the error of the rifle, and gives information as to any constant fault of the firer.

379. Should a soldier make a bad group, and the rifle is suspected, it should be at once fired under similar conditions by a reliable marksman. Should the marksman also make a bad group with the rifle, the rifle should be tested under paras. **125** and **126**, and if found "inaccurate," the man to whom the rifle belongs should be allowed to recommence the course, the necessary ammunition being found from the present authorized allowances. If the rifle is proved accurate, the soldier's aim should next be tested by the triangular method, and his let-off should be tested by means of the aim corrector. His eyesight and nervous condition should be examined.

380. Unsteadiness may be traced to lack of determination, or to illness, or to some habitual excess, such as cigarette smoking. The position of the shots on the target may indicate errors in aiming or inclination of the sights; they may show that the firer moves his shoulder forward to meet the shock of discharge, which will tend to throw shots low left, or that he flinches from the shock, directing them high. If the trigger is jerked by the wrist instead of being pressed by an independent action of the finger, shots will usually strike low right.

381. Instructors should carefully note the positions of good groups as well as bad ones, for some constant error in aiming of fault of the rifle may thus be discovered, which will escape notice in application practice when every shot is signalled and error is attributed to wind or other cause for which allowance is easily made in sighting. Such errors are not uncommon, even among marksmen, and often affect their shooting unfavourably in field practices when there is no shot for shot signalling.

382. Some men can aim quickly but lose their power of definition when straining the eye. In serious cases, the soldier should be medically examined with a view to discharge or the provision of proper glasses.

383. The will-power of men who have shown nervousness should be developed by gymnastics or other means. Faults may sometimes be remedied at once by correcting the fire positions, by allowing time to elapse between the shots or by snapping practice. Sometimes the cause of the failure is due to the firer's effort to succeed ; he dwells too long on his aim, his muscles become tired, the vision becomes blurred and his will-power is impaired. It is necessary to take time in such cases.

384. A complete analysis of the faults of the firer and his rifle should be made before leaving the range, and a note should be made on the register of the steps decided upon for remedying defects.

385. A sub-target machine may be used with advantage for the analysis of faults.

Section 78.—*Application.*

386. When a recruit is shown by the result of his grouping to have acquired sufficient skill in aiming and trigger pressing, he will be instructed in applying his shots to a definite mark. This form of practice, in which aim or sighting is altered as may be found necessary, is called " Application."

387. Instruction is given in making allowance for atmospheric influences, chiefly cross-winds ; but it is not desirable that trifling changes of wind should be met by minute adjustment of the windgauge. The instructor should call upon the soldier to estimate the wind before firing, and tell him the corrected allowance which he is to place on his sights. Subsequent alteration should be unnecessary. Similarly, small errors in elevation should be met by aiming up and down, and not by fine adjustment, though the target must always be kept in view.

388. Grouping standards are in some cases attached to application practices in order to emphasize the importance of care and consistency in shooting.

Section 79.—*Special instructions.*

389. It is intended that the standards fixed for qualifying practices shall ensure the allotment of extra ammunition to inferior shots.

390. Instruction on the firing point is an indispensable form of musketry instruction for young soldiers, but if it leads to continual alterations of sighting to meet errors in shooting, the firer is confirmed in his errors and his faults are only obscured. During the firing the instructor should watch the recruit, not the target, and should insist on being told the probable result of the shot before it is signalled.

391. No departure from correct firing positions should on any account be permitted ; the rifle must be gripped, the face kept back from the right hand, and there should be no constraint.

392. The management of the breathing and the let-off must be noticed and the recruit reminded of them continually, so that his mind may be centred on the more important details of shooting and not on changes of wind or light, with which he will become familiar later.

393. Although care and deliberation are necessary in elementary firing instruction recruits must not be allowed to fall into the habit of dwelling on their aim nor of aiming and returning to the loading position repeatedly before pressing the trigger. These errors arise chiefly from taking a fine sight and accommodating the eye so as to obtain a defined image of the foresight rather than the target. When such methods are adopted it is a sign that the object of range practices is misunderstood, and that the firer is in need of more practical instruction.

394. There is no object in establishing a phenomenal standard of accuracy in elementary range practices and deliberate shooting; a satisfactory degree of proficiency is soon attained by the majority of men and they should then proceed to timed instructional practices. It is in timed shooting at distances below 600 yards that a very high degree of proficiency is desired.

395. In snapshooting practices, not only the exposure of the target but also that of the firer should be limited; strict attention must be paid to positions behind cover, and to firing with the least possible movement and exposure.

396. Snapshooting should be freely practised on 30-yards and miniature ranges, so that any tendency to jerk at the trigger may be overcome by constant practice. Rapid grouping may be practised similarly with a sub-target machine.

397. In rapid firing a man of normal temperament should be able to attain the regulation rate with trifling loss of accuracy, but it is not desirable to make a great sacrifice of accuracy to produce even the regulation rate. Dexterity in loading and a habit of rapid alignment of the sights should be developed in preliminary training ; in the range practices the opportunity is afforded to every man to ascertain his own best rate for combining accuracy in shooting with rapidity of fire so as to produce a high average of hits per minute, but there is no obligation to fire all the rounds allotted in any rapid practice.

398. The regulations for repetition and instructional firing are intended as a guide to officers responsible for instruction, and need not be observed in the letter except in so far as they govern qualification* or classification, when strict observance of all conditions is necessary.

399. Every soldier should keep a record of the results of each practice, together with the prevailing conditions of light and atmosphere, and the sighting elevation which he found to be necessary. The cause of any failures should be noted. Figures of tests in judging distances and preliminary training should also be recorded.

400. Boys will be trained in the use of the rifle at the discretion of their commanding officers.

---

* The word qualification in these Regulations means that the necessary standard for advancement to a further course of instruction has been attained. It is not to be taken as meaning qualification for pay, or for earning grants or allowances.

Section 80.—*Recruits' Course, Regular Forces, Cavalry, R.E., and Infantry.*

401. Part I should be fired intermittently during the latter part of preliminary training and may be repeated as often as is considered necessary (*see* para. **136**). The conditions may be varied with the object of removing any particular defects observed in previous shooting.

402. On the completion of Part II recruits will repeat once those practices of Part II in which they failed to reach the Grouping Standard, or if they obtain less than 90 points in Part II they will repeat once the whole of Part II before proceeding to Parts III to VI.

403. Recruits who have completed their course of instruction in range practices, fire elementary field practices before joining their companies in the trained soldiers' course, in order that they may learn to regard elementary shooting as a means to an end.

Section 81.—*Execution of Tables A and B in the same year.*

404. Soldiers of the cavalry, Royal Engineers, and infantry of the regular forces who complete Table "A" will be exercised in the whole of Table "B' (sappers of the Royal Engineers the prescribed practices only). with their own companies, if possible, in the same year, but their scores in Part III. will not be included in the company or battalion averages. In special cases, which should be exceptional, commanding officers may direct that backward men shall repeat Part II., Table "A," instead of firing any portion of Table "B" in the same year.

405. If Part II., Table "B," has been commenced before they are available, they will begin firing at any practice which their companies are executing at the time they become qualified, and will then complete the remainder of the Table.

If all the companies in the battalion have completed Part II., Table "B", before they are available, the brigade commander will decide as to whether they shall be exercised as casuals, *vide* para. 413.

406. They will be awarded marksmen's badges if they attain the necessary standard in Table B.

407. In any case they will receive such further practice in firing as may be considered by their company commanders to be necessary (para. 139).

TRAINED SOLDIERS' COURSES, REGULAR FORCES.

Section 82.—*General.*

408. Every soldier not exempted by these regulations or by the "King's Regulations" will execute the full course of range and field practices yearly.

409. Subaltern officers will fire the range practices of Table B with their companies. It is optional for company commanders and for officers, other than company officers below field rank doing duty with a unit, to do so. Officers will fire with the weapon with which their unit is armed.

410. Bandsmen and signallers may fire either with their companies or in separate parties. In the latter case their performances will be recorded in the same manner as those of casuals. Machine-gun sections will fire together with one company.

411. If a man has fired one or more rounds in any range practice and is prevented from completing it, the points made will not count, and the whole practice will be re-commenced when his training is resumed.

412. Warrant officers, staff serjeants, serjeant pipers, warrant and non-commissioned officers of the Army Gymnastic Staff, those who under normal conditions would not use the rifle in war, and those who in war would not be in command of men who use the rifle, provided they have once been fully exercised, will be exempt from musketry practice unless the General Officer Commanding otherwise orders; or unless the issue of their proficiency pay is dependent on musketry qualification. They will be classified according to the standard obtained when they last fired.

The following will be partially exempted:—

i. Bandsmen, pioneers, trumpeters, buglers, drummers, and men of the machine gun detachment will fire the range practices of Table B only.

ii. Non-commissioned officers and men appointed to "K" Signal Company and to the Survey Companies Royal Engineers will be exercised biennially.

iii. Qualified signallers up to the number authorized in establishments will be exempted from collective field practices.

413. All men not fully exercised in Table "B" with their companies (except those referred to in the first sentence of para. **405**) will be attached to other companies to carry out the range or field practices omitted; or, if all companies have completed Table "B", a party of casuals may be formed to ensure that all men qualified are fully exercised in the whole of Table "B."

414. The commanding officer will apply that a soldier detached from his unit may be exercised, at the station where he is serving, with a unit of his own branch of the service.

Section 83.—*Special instructions.*

415. Range practices are fired by trained soldiers in order that they may revise their knowledge of elementary and timed shooting before entering upon more advanced practices.

416. Soldiers who have missed the whole or a portion of the range practices and have become available to commence the field practices with their companies may, if they are known to be good shots, be allowed by their commanding officers to execute the field practices, and fire the range practices subsequently.

417. Considerable latitude is allowed as regards the programme of instructional practices. Officers commanding companies may vary the number of rounds to be fired by individuals, or they may alter the order of the practices in Part II. Officers commanding battalions may, with the approval of general officers commanding brigades, vary the instructional practices in any way calculated to further instruction, but it is not permitted to design practices, or to vary the details of practices in Part II with the object of assimilating the conditions to those of the classification practices.

418. Correction of sighting in individual firing is rarely possible in war. It is therefore all-important to estimate the elevation and deflection for the first shot. When a reasonable standard of skill in trigger pressing has been shown in grouping practices, and the principle of application is understood, further practice in slow fire should aim at successful application of fire from the first shot, and less importance should be attached to correction of sighting according to the signalling of a series of shots.

For this reason skilled shots should fire two or three shots at each of several ranges for sighting practice rather than long series of shots at one or two distances. Only a few rounds in all should be devoted in their case to deliberate shooting; a high standard of snapshooting should be developed. Officers and serjeants may fire sighting shots at ranges beyond 600 yards, but as a rule such training should be reserved for the fire direction practices. It is convenient to memorize the effect of right angle winds at some one distance, as a guide in estimating deflection allowance for winds of similar strength at other distances. 500 yards is a satisfactory range for this purpose, and the approximate effect of right angle winds blowing 10, 20, and 30 miles per hour may be studied with advantage; the use of elaborate wind tables and dependence on flags, telescopes, or sighting shots is prohibited.

419. The change which takes place in conditions of shooting when there is no signalling of each shot is not always appreciated, and it is essential that the importance of the first application of fire should be realized. Troops, therefore, which are unable to fire field practices should fire some at least of the instructional practices without signalling. Such practices are called observation practices; the targets should be placed on the face of the stop butt or at the foot of the gallery bank, and may with advantage be falling or collapsible targets.

420. Special reports will be submitted in the case of men who are found to be physically incapable of reaching the 2nd Class standard in Table B.

Section 84.—*Allotment of ammunition and computation of averages.*

421. The allotment of ammunition will be as follows:—

Part I.	Part II.	Part III.	Part IV.	Part VI.	Surplus.
30	60	50	35	50	25

Total rounds = 250.

422. Ammunition required for repetition of qualifying practices will, when necessary, be taken from the surplus.

423. The best shot of each squadron or company will be determined by his aggregate score in Part III.

424. Averages made by companies in each of the classification practices will be calculated to one place of decimals and published in regimental orders. Only the scores of those officers, non-commissioned officers, and men who completed a practice will be included in the numbers by which the total points made in that practice are divided. Points made by casuals will be included at the end of the year and fresh averages struck. The sum of the averages of the practices of Part III will be termed the company "average" in classification practices.

Section 85.—Conditions of qualification.

425. The conditions for qualification in Part I, Table B, are:—

A score of not less than 15 in each of Practices 1 and 4, and a total of not less than 45 in Practices 2, 3, 5 and 6.

426. Those who fail to reach any of these standards will, after firing Practices 7 to 14, repeat those practices of Part I in which they failed, until the standard is attained or a third failure is recorded. They will omit Practices 15, 16, and 17. Those who fail to reach all the standards after two repetitions of those practices in which they failed, will be classified as 3rd Class shots, and will not fire Part III, but will fire Part VI if ammunition is available.

Section 86.—Classification practices and conditions of classification.

427. The classification practices should be fired during the most favourable time of year for individual firing. As a man's pay is affected by these practices, every facility should be given for firing to be carried out under favourable conditions.

428. With the exception of those who are exempted from musketry by the regulations, every man on the strength of his unit on the last of the four days allotted to classification firing, will be classified at or before the end of the year.

429. Trained men, not exempted by the regulations, who for any reason do not commence Table B will be classified as 3rd Class shots. Those who commence Table B, Part III, but do not complete it, will be classified according to the number of points obtained in those practices which they complete.

430. A note will be made in the Company and Battalion annual returns of the number of men (if any) who are classified without having completed the range practices of Table B.

431. Cavalry and infantry soldiers, if qualified in Part I, will be classified upon their total scores in Part III as follows:—

Those who obtain 130 points and upwards...	Marksmen.
" " 105 " " less than 130 ...	1st Class shots.
" " 50 " in Practices 18 to 22 inclusive, or in the alternative obtain 70 in Part III	2nd Class shots.
" for any reason fail to attain the 2nd Class Standard	3rd Class shots.

432. Commanding officers will take every opportunity of stimulating all ranks to take an interest in shooting and judging distance by granting indulgences to the most proficient and by giving prominence to the best shooting company in any manner considered desirable.

433. Third Class shots will not be employed in any capacity which will interfere with their attendance at all parades and instructional duties.

Section 87.—*Royal Engineers.*

434. The Sappers and Pioneers of the Royal Engineers (Regular Forces)* will fire the following practices :—

i. *Table A.—Recruits' Course Regular Forces* (*Cavalry, R.E., Infantry*), *and Special Reserve* (*R.E., and Infantry*). Parts I, II, III, and IV.

They will not repeat Part I, nor any practices of Part II, unless they fail to make a total score of 90 points in these practices.

Those who fail to obtain 90 points will, after firing Part III, repeat practices 5 to 12, instead of firing Part IV.

Surplus ammunition will be used for the further training of indifferent shots, for testing rifles, or for occasional shots.

Table B. Annual Course, Regular Forces (*Cavalry, R.E., and Infantry*). *Parts I., II.,* (*Practices* 7, 9, 12 *and* 14), *and III.* The qualification standard will be the same as for the cavalry and infantry. Those who fail to qualify in Part I. at the first attempt will proceed with the prescribed practices of Part II. (20 rounds) and then repeat those practices of Part I. in which they failed. Those who fail twice in any of the standards in Part I. may, provided sufficient ammunition is available out of the authorized annual allowance, repeat those practices in which they failed once more, with a view to qualification. Those who

* The musketry courses prescribed for recruits R.E. drivers are contained in Appendix II, those for trained R.E. drivers in Appendix I.

repeat any practices of Part I. will, if they qualify on first or second repetition, fire the whole of Part III., or, failing that, Practices 19, 20, 23, 24 and 25, according to the amount of ammunition available.

Any man who qualifies in Part I. on the first, second, or third attempt, and completes the practices of Part III., will be classified according to the standards of Part III., but not below 2nd Class shot. Those who complete practices 19, 20, 23, 24 and 25 only will be classified as 2nd Class shots, provided they have qualified in Part I.; those who fail in Part I., or who, having qualified in Part I. for any reason do not complete these five practices will be classified as 3rd Class shots.

Any ammunition unexpended will form a surplus for further instruction.

Section 87A.—*R.A.M.C. and A.V.C.*

434A. Recruits of the R.A.M.C. and A.V.C. will fire practices I to IV, Table B (Royal Artillery, &c.), Appendix II. The following standard will be required :—

Practice 1.—All shots in a 12-inch ring. Those who fail to attain this standard will repeat Practice 1 until they do so, or expend the full allowance of ammunition. Ammunition not required for the above practices is to be expended on preliminary training on the 30 yards range as considered necessary.

Trained soldiers of the R.A.M.C. and A.V.C. quartered in South Africa will fire Table B (Royal Artillery, etc.), Appendix II.

THE CONDUCT OF RANGE PRACTICES.

Section 88.—*General instructions.*

435. General officers commanding will cause such orders to be framed as local conditions necessitate to ensure safety, and will arrange that copies are always available for reference during practice.

These orders will contain information as to the range duties required on ranges of large extent, used by several units at one time. They should be as brief as possible, and should contain no amplification or repetition of instructions laid down in the regulations. The duties of range wardens should be detailed. Certificates of handing and taking over stores for daily use and return should be unnecessary.

436. Range practices, unless otherwise ordered, will be fired in drill order.

437. Range practices should be fired, as far as possible, in favourable weather. It is of the utmost importance that recruits' firing should not take place in cold and unsuitable weather.

438. Range practices should, as a rule, be fired in the order in which they appear in the tables, but brigade commanders may vary the order at their discretion. When deliberate and rapid practices for classification are fired at the same distance, each man may fire the rapid practice immediately after the deliberate practice.

439. In the case of ranges of less than full extent, general officers commanding-in-chief may frame special instructions, make proportionate changes in the size of targets, and vary the points for classification; but the general distribution of ammunition laid down in the several parts of Tables A and B respectively should be adhered to.

440. A record of such variations will be forwarded to the commandant, School of Musketry, for the information of the Army Council.

441. In the Regular Army and Special Reserve not more than 15 rounds should be fired in one day except in classification practices, when 25 rounds may be fired if necessary.

442. It is always better, when time is pressing, to reduce the number of rounds fired in instructional practices than to hurry through them. All available officers should be present at the firing points during instructional practices.

Section 89.—*Special instructions.*

443. *Grouping practices.*—i. One firer will be detailed to each target and fire five shots, maintaining the regulation point of aim throughout. Targets will be changed, and a second detail of men will fire similarly.

ii. Both details will then proceed to the targets, see their groups measured, and note the positions of the points of mean impart with reference to the points aimed at. If it is impracticable to proceed to the targets, the group may be marked by means of small spotting discs.

iii. The groups will be measured with wire rings, 4, 8, and 12 inches in diameter, counting 25, 20, and 15 points respectively; 10 points will be allowed for a 12-inch group with one wide shot.

iv. The ring which will contain all the shots will be recorded as the measure of the group. A shot mark is included within a ring when it cuts the circumference of the largest circle which can be described within that ring by means of a pencil held at right angles to the target.

v. All shot marks found on a target will be included in the group to be measured. No points will be allotted to a group unless there are five shot marks at least on the target. If more than five shot marks are found on the target there will be no score, and the practice will be repeated.

vi. When the ring is placed to include all the shots, the centre of the ring will be taken as approximately the point of mean impact; its distance from, and direction with reference to, the point aimed at will be recorded on the register (*e.g.*, 7 inches, 4 o'clock).

vii. On return to the firing point other details will fire, but steps will be immediately taken to ascertain the cause of any bad shooting of men in the first two details.

444. As a rule 3rd Class shots should not be allowed ammunition for further training in application practice until a satisfactory standard in grouping has been attained. This may be fixed at a ring of diameter equal to $\frac{1}{300}$th of the range, but officers should exercise their discretion as to allowing one wide shot in five when dealing with young soldiers.

445. In slow practices 20 seconds is the time limit allowed for each shot, reckoned from the act of loading. If there is a tendency to exceed the limit, a whistle should be used to mark the beginning and end of each period, but not otherwise.

446. In firing from behind cover, the position adopted must be such as would enable the firer on service to obtain the fullest protection from the cover, having due regard to the efficiency of his fire. In the prone position the grip of the left hand must be maintained on the rifle, and there must be no undue exposure of the shoulder or legs.

447. The timed exposure of targets for snapshooting and rapid fire practices will be reckoned from the time when the target is in position and stationary, to the time when it is again moved for lowering. The movements of raising and lowering must be conducted with the utmost rapidity, but without jarring the target frames.

448. Timing in rapid practices should be reckoned from the word of command "*Rapid Fire,*" and fire should be stopped by the command "*Cease Fire.*" The command "*Rapid Fire*" should be given as soon as the target appears. The target should be lowered to half-mast at the end of the time allowed for firing under orders of the officer on butt duty, but the officer superintending at the firing-point should also time the practice and order "*Cease Fire*" at the end of the time allowed for firing, reckoned from the command "*Rapid Fire.*" Four points (3 points in practices in which the highest possible score is 3 per hit) will be deducted for every shot fired after the order to cease fire has been given.

449. In firing from behind cover the butt of the rifle will be in contact with the ground, and the firer will remain in observation, but otherwise completely covered, until the command "*Rapid Fire*" is given in rapid firing practices, or the target appears in "Snapshooting" and "Crossing Shot" practices. When snapshooting, or firing rapid in the open, the rifle may be held in the loading or aiming position as preferred.

450. In rapid practices, unless otherwise stated in the Instructions for the Conduct of the Practice, the magazine will be charged with 4 rounds, and the rifle will be loaded before the target appears.

451. No instruction or assistance of any kind will be given to any man during the firing of Part III, Table B (Cavalry, R.E., and Infantry).

452. The use of the windgauge or of the fine adjustment will not be permitted in any classification practice or standard test practice.

453. In the event of a jamb occurring in a timed practice, and provided that it is not caused by any fault on the part of the firer, the time allowed for the practice will be increased to the extent due to the delay caused thereby. Should, however, a jamb in a rapid practice be due to a breakage of mechanism or other defect that cannot readily be rectified on the range, the whole practice will be fired again.

In the event of missfires, extra rounds will be allowed equal to the number of missfires in the practice concerned, a proportionate part of the time allowed for the whole practice being given for each extra round.

Whenever extra time is allowed for a timed practice, a report giving the reason, and stating whether the jamb or missfire was due to the rifle or to the ammunition, will be rendered to Command Headquarters.

454. Omission to fire the rounds allotted and failure to fire during an exposure or run in vanishing and moving practices will entail forfeiture of the rounds that should have been fired, and misses will be recorded for them.

455. In range practices the regulation positions are obligatory, except in firing from cover, when the rifle should be rested and the position adapted to the ground. This includes resting the arm as well, if suitable, but the cover is not to be specially constructed as a rest for the rifle.

456. Loading will always be through the magazine.

457. Dependence on the sling should be discouraged, and it will not be used for steadying the rifle in range practices.

458. No soldier equipped with a rifle is permitted to fire with any but his own.

459. No sighting shots are allowed.

460. Occasional shots to verify elevation or strength of wind, or to prove the accuracy of a rifle, may sometimes be fired by an officer or non-commissioned officer, with the senior officer's permission. They will not be fired during classification practices or standard tests. Notification of their commencement and conclusion will be made to the officer in the butts by telephone, signal, or bugle sound. The target in use will be lowered and checked, and a clean one raised for the occasional shots. When they are completed, it will be lowered and checked, and the original target raised for the firer to complete his rounds.

461. In order to ensure that full effect is given to the directions contained in para. **390**, men will fire singly, never in twos or threes.

462. A coat or waterproof sheet may be used to protect the uniform, but except when firing from cover or when rests are authorized, neither rifle, forearm, wrist nor hand is to rest against anything or to be supported.

463. No man will load, or assume a firing position, until the senior officer present has ordered the practice to commence. After firing, men will return to the loading position, but will not open the breech in the *slow* practices until the last shot has been signalled. If it is necessary to suspend firing, all men who are in position will raise the safety catch (or unload if no safety catch is provided), until the order is given to resume the practice.

464. Aiming or snapping during target practice may only take place from the firing point after the red flag has been lowered.

465. No one is allowed at the firing point except the men actually firing, the instructors, and officers. All non-commissioned officers and men not on duty at the firing point will ground or pile arms and remain not less than 30 yards in rear of the firing point.

466. During intervals of firing an opportunity should be given occasionally to all ranks for revising their impressions as to the visibility of the human figure at short distances, by placing men on the firing platforms up the range. The study of visibility under conditions of known distance and in relation to targets used for practice in shooting is especially valuable.

467. Those men who are not actually engaged in firing should receive instruction in ground reconnaissance, use of the eyes, use of field glasses, range finding and description of ground, while waiting behind the firing point.

468. No shouting is allowed; men next to fire will be brought up by signal.

469. Field glasses or telescopes will be carried by all officers and section commanders. The men should be encouraged to use them during spare time on the range.

470. The sights will be used as issued, without alteration of any kind. They may not be blackened; the browning is renewed by an armourer when required. No additions, marking or colouring are permitted, nor are orthoptics allowed.

Section 90.—*Webley pistol practice.*

471. Blank firing will be carried out as follows :—

i. By cavalry, before firing the mounted practice with ball—

10 rounds single practice, mounted.
10 rounds continuous practice, mounted.

ii. Others for whom blank ammunition is allowed—

5 rounds single practice.
5 rounds continuous practice

472. Ball firing will be performed in the following manner, each practice consisting of 6 rounds :—

i. *First year.*

		Rounds.
1st Day.	One single practice with each hand, dismounted	12
2nd Day.	" " " "	12
3rd Day.	One continuous practice with each hand, "	12
4th Day.	" " " "	12
5th Day.	Two single and two continuous practices with the right hand, mounted	24
6th Day.	" " " "	24
	Total ...	96

Officers of dismounted units will not fire the mounted practices, but will repeat the practices of the first 4 days.

ii. *Subsequent years.*

		Rounds.
1st Day.	One single practice and one continuous practice with each hand, dismounted ...	24
2nd Day.	Two single and two continuous practices with the right hand, mounted ...	24
3rd Day.	Two single and two continuous practices with the right hand, mounted ...	24
	Total ...	72

Officers of dismounted units will repeat the practices of the first day once only instead of firing the mounted practices.

iii. The days should not be consecutive.

473. Every possible precaution is to be taken against accidents, the strictest order and discipline being maintained at the firing point.

474. The practice will be conducted as follows, viz.:—

i. Each man will be individually called to the firing point by the superintending officer; he will load and fire two practices, and will then return to his original position before the next man is called up.

ii. A second class elementary (bull's-eye) target will be used, but commanding officers are authorized to adopt any other form of target which they may consider more suitable, and to vary the scoring.

iii. 30 yards ranges may be used.

iv. The rounds will be fired in the standing position at 30 yards range.

475. i. The scoring, if regulation targets are used, will be as follows, viz.:—Bull's-eye, 4; inner, 3; outer, 2.

ii. The scores will be recorded on the register on the conclusion of each practice.

iii. Army Form B 190A, will be the register used.

iv. The numbers exercised will be shown in the annual musketry return.

476. All surplus ammunition will be disposed of under the orders of the commanding officer.

Section 91.—*Range duties.*

477. i. During the following practices, which are inspection tests of the results of training, officers, non-commissioned officers and men from units other than that which is firing will be detailed under the orders of general officers commanding, for all duties of supervising and marking, except in the Territorial Force and in the infantry of the Special Reserve, where the independent supervision will be carried out in the butts only, either by officers of other corps or, when this is impracticable, by officers of other companies of the same unit. In the infantry of the Special Reserve, and in the Territorial Force, instruction may be given to non-commissioned officers and men during the practices of Part II, Table B (Appendix III), and of Tables A and B (Appendix IV) :—

Table B. Regular Forces (Cavalry, &c.) Part III. (By whomsoever fired.)
" B. Special Reserve. Part II.
" A. Territorial Force. Part II.
" B. Territorial Force. Part II.
" C. Regular Cavalry and Infantry, Part II. Classification practices 9, 10, 11 and 12

ii. For the above practices the provisions of paras. **483**, **484**, and **485** will be scrupulously carried out.

iii. If markers cannot be found from other units, officers of other units, at least, should be specially detailed for supervision at the butts and the firing point.

iv. Officers of the squadron or company which is firing will be detailed to assist superintending officers at the firing points as may be necessary. In rapid practices their special duty will be to report on any jambs that may occur.

478. For the undermentioned practices, there must be equally thorough supervision by officers at the butts and firing point, but they may in this case belong to the unit which is firing :—

Table B.	Regular Forces (Cavalry, &c.).	Part I.
„ A.	„ „ „	„ Part IV, when fired by recruits of the Special Reserve.
„ B.	Special Reserve.	Part I.
„ B.	Royal Artillery, A.S.C. and A.O. Corps.	

479. The duties specified in para. **485** will be carried out during practices other than those mentioned in para. **477** by non-commissioned officers under such supervision as may be considered necessary.

480. To guard against accidents, the following orders will be observed :—

i. No firing will take place until a large red danger flag is hoisted on the signal staff at or near the butts, and the necessary look-out men posted.

ii. A smaller red danger flag will be hoisted at the butts as a warning to cease fire. This flag will remain exposed during the entire period of cessation of fire, and will not be withdrawn until the whole of the butt party is under cover. No one will leave the butts until cessation of fire has been notified from the firing point.

iii. A red flag will be kept raised at the firing point when no firing is taking place, and will be lowered only on the order of the senior officer. This order will not be given until the flag at the butts has been withdrawn.

iv. When cessation of fire is required, the senior officer at the firing point will give the order. When all fire has ceased, he will cause the red flag to be raised, and the butt party to be informed.

481. Permanent fatigue parties should not be required; companies at practice have usually time to perform any necessary work. Minor fatigues, such as hoisting flags, drawing and issuing stores, &c., should be carried out by the range warden, &c., permanently employed on a range.

482. It is not possible for an officer to adequately supervise more than four firing points. Companies and other parties should not therefore, in ordinary circumstances, be allotted more than four targets.

483. The duties of the officer superintending at the firing point on a gallery range are :—

i. In the practices mentioned in paras. **477** and **478** to vary the order of firing before the commencement of each practice.

ii. To see that each man on his roll (Army Form B 189) fires at the target to which he has been detailed, and fires with the rifle which has been issued to him. In the practices of Part III., Table "B," the number of the rifle used will be entered on Army Form B 189.

iii. During preliminary and instructional practices to allow no person at the firing point but the officers, the instructors, and the men actually firing.

iv. To ensure that the regulations as to target practice and local orders are obeyed.

v. To detail a non-commissioned officer or man to send and receive messages on the telephone, or by signal.

vi. During preliminary and instructional practices to detail a non-commissioned officer to superintend each man's firing, and to instruct. (*See* paras. **390** and **461.**)

vii. To see that no assistance by way of coaching or spotting is given to non-commissioned officers or men while firing the practices of Part III, Table B (Cavalry, R.E., and Infantry).

viii. To ensure that no more than the authorized amount of ammunition is expended.

ix. To collect the butt registers from the officers on butt duty. These registers will be handed in to the headquarters of the unit finding the supervising officer or to the office of the Camp or Station Commandant, where they will be stamped and forwarded to the unit to which the firing party belongs.

x. To see that no two firers use the same rifle (officers and others not issued with a rifle excepted) and to record the numbers of the rifles used on the grouping registers.

xi. To ensure as far as possible, by personal observation, that the marking in the butts is carried out in accordance with the regulations.

xii. For other instructions, *see* Section 89.

484. For the practices referred to in paras. **477** and **478** officers will be detailed to check the hits and supervise marking in the butts in the proportion of one officer to a group of not more than four targets.

For Part II, Table C (*see* paras. **646** and **647**), an officer, not necessarily a machine gun section officer, will be detailed for butt duty to check hits, to ensure that they are correctly entered, to see that the vertical lines for practices 10 and 12 are correctly drawn on the targets, and to supervise generally. His duties will also include those enumerated in para. **485**, i, ii, iii, iv, ix, xi and xiii.

485. The duties of an officer on butt duty in a gallery range are as follows :—

i. To see that the targets are of the proper dimensions, and sufficiently clean to enable shot holes to be easily distinguished.

ii. To see that the butts and appliances are in good order, and to report damage and deficiency.

iii. To explain all regulations and local orders to the markers and to ensure their observance.

iv. To detail markers to targets.

v. To ensure that no target is lowered without his express permission. (In slow practices the target will not be lowered at all until the officer is in front of it. In rapid practices the target will be lowered to half-mast at the end of the time allowed, and the markers ordered to stand as far back as possible until the officer is in front of it.)

vi. To cause all targets to be lowered during cessation of fire.

vii. To regulate the exposure of moving and vanishing targets according to the instructions laid down.

viii. To see that marking is accurately performed.

ix. To personally check the target of each firer and enter

the value of all hits in the register (A. F B 190, or B 2050A), in ink; occasional shots will be entered in the columns provided for the purpose. No erasure is to be made. If alteration is necessary a fine line will be drawn through the figure, the correct value written against it, and the amendment verified by the officer's initials.

x. If more hits, including ricochets, are found on a target than rounds fired, to deduct from the score the value of the highest hits. Only those hits which are to count are to be entered on the register.

xi. To mark off each hit on the target, with a *red* pencil, before entering its value in the register, and to ensure that all shot holes are duly patched.

xii. In rapid practices, after each check, to cause the number of hits of each value to be signalled on each target.

xiii. On the conclusion of a practice, to rule a line diagonally across the unused spaces in the register, before signing it.

486. A non-commissioned officer not under the rank of corporal will be detailed to assist each officer on butt duty.

487. Two markers will be allotted to each target.

488. i. *On non-gallery rifle ranges*, or in special cases at the discretion of the commanding officer, the register (Army Form B 190A), on which all hits or misses will at once be entered as signalled, will be kept at each firing point by a non-commissioned officer specially detailed for the purpose.

ii. Targets will only be checked on the completion of the rounds allotted in timed practices, after occasional shots, or when the number of hits renders marking difficult. The officer on butt duty will signal the warning to cease fire, examine targets, mark off all hits or ricochets, and enter their value on the memorandum (Army Form B 67) supplied for the purpose. He will then cause them to be signalled to the officer at the firing point (a marking disc showing the value of each series being placed on the target), who will compare them with the numbers recorded on the register, and enter the totals in the spaces provided for the purpose. The hits will then be patched.

iii. In other respects the regulations for gallery ranges will be observed.

489. The officer commanding a company or party is responsible that arms and pouches are examined before men leave the firing point.

Section 92.—*Signalling and scoring.*

490. Signals between the butts and firing point will be made by means of a flag as follows :—

(The signaller, whether at the firing point or butts, will face the target.)

Preparative.	Waved above the head.
Ready to receive or send down.	Held vertically above the head.
Tens.	Lowered to signallers' right as many times as required.
Units.	Lowered to signaller's left as many times as required.
No hits.	Revolved two or three times from right to left.
Wash (*or patch*) *out.*	Waved horizontally close to the ground.

Practice finished. Held upside down.
Signal last shot. Held horizontally above the head.

491. The following code is suggested as suitable for use on ranges where communication between the firing point and butts is maintained by means of signals observed in reflecting mirrors.*

i. The signaller, provided with a large signalling flag, should take up his position exactly opposite to and facing the mirror, and his signals would refer only to the four targets on which the party is firing, numbered from the right when facing the firing point, irrespective of their numbers with reference to other targets on the range.

ii. Signals should be acknowledged from the butts by means of a small signalling flag raised behind the mirror.

No.	Flag Signals.	Their significance
1	"Preparative," followed by flag waved vertically up and down in front of the body.	The practice is about to commence.
2	"Preparative" flag raised to full extent of right arm, and revolved from left to right in front of the body.	Occasional shots are about to be fired.
3	As for No. 2, followed by "wash out."	Occasional shots are finished.
4	As for No. 2, followed by signal for target.	Occasional shots are about to be fired on No. target.

* *See* Part II of these Regulations.

No.	Flag Signals.	Their significance.
5	As for No. 4, followed by "wash out."	Occasional shots are finished on No. target.
6	Make the "preparative," and hold the flag vertically above the head to the full extent of the left arm. Then lower it to the left as follows:— For No. 1 target, once. ,, No. 2 ,, twice. ,, No. 3 ,, three times. ,, No. 4 ,, four times.	To call up a target.
7	As for No. 6, followed by "wash out."	Check No. target. NOTE.—To check all targets, hold the flag horizontally to the right followed by "wash out." This signal can be repeated if it is necessary to re-examine.
8	"Preparative," followed by flag held upside down, with bunting spread out.	The practice is finished.
9	As for No. 8, followed by flag circled above the head.	Practice is concluded for the day.

432. In the practices specified in paras. 477 and 478, targets, except in timed practices, when they will be brought to half-mast, will only be lowered on a direct order from the officer on butt duty, but every shot from the firing point will at once be signalled at the butts, as a hit, ricochet, or miss. In timed and observation practices, the number of hits of each value will be signalled at the expiration of the time limit. In timed practices, after the targets have been checked.

493. In slow practices, when a target is struck, the marking disc will be placed on the shot hole and kept in position sufficiently long to enable the firer to see the position of his shot.

494. When a shot strikes the target so that the circumference of the mark cuts the outer edge of any ring or figure it is to be counted as hitting within that ring or figure as the case may be. No shot is to be counted unless the whole or part of the mark of the bullet is seen on the face of the target. A *ricochet* usually makes a long ragged hole or mark.

495. For elementary and figure targets. For scoring in Practice 22, Table B, Appendix I, *see* Note 2, below.

Signal.	Methods of signalling.	Value of hit.
Bull's-eye or figure. For Practice 22, Table B, Appendix I, *see* Note 2, below	Polished metal or white disc placed on shot hole.	4 points.
Inner (remainder of inner circle)	Black disc waved twice across the face of the target and placed on the shot hole.	3 "
Outer (remainder of elementary target) or magpie (remainder of large circle on figure target)	Polished metal or white disc revolved in front of the target and then placed on the shot hole.	2 "
Outer (remainder of figure target)	Black disc moved vertically up and down the left of the target and then placed on the shot hole.	1 "

Signal.	Methods of signalling.	Value of hit.
Ricochet or miss ...	Red and white flag shown on the same side as the direction of the miss. If the direction cannot be determined the flag will be waved across the face of the target.	Nil.

NOTE 1.—The scoring bull's eye on 2nd class elementary targets is a 12-inch invisible ring.

NOTE 2.—In Practice 22, Table B, Appendix I, a hit on the figure or remainder of inner circle will count 3 points and will be signalled as a bull's-eye. Hits elsewhere on the target will count as in table above.

NOTE 3.—When for any reason it is found to be impracticable to send the firers into the gallery after a grouping practice the following signals may be used :—

Bull's-eye signal denotes a 4-inch group.
Inner ,, ,, 8 ,, ,,
Magpie ,, ,, 12 ,, ,,
Outer ,, ,, 12 ,, ,, , with one wide shot.

When the signal has been made, after a short pause, the point of the pole will be placed on the point of mean impact of the group.

496. For snapshooting practices the 2nd class figure target will be used, but all hits outside the magpie circle will be signalled as misses, and will not be shown on the register. The method of signalling and value of hits will otherwise be the same as in para. **495.**

497. In observation practices hits only will be signalled by raising a marking disc in the markers' gallery. Collapsible targets should be used if available. Collapsible targets, if used in snapshooting practices, must also be vanishing targets. If iron targets are used, care will be taken that they must fall and cannot turn when hit.

498. *On figures, No. 3 and No. 6.*

Hit	The figure will be raised above the markers' gallery and twirled.	3 points.
Ricochet / Miss ...	As in para. **495**	Nil.

499. In snapshooting practices, and in practices at moving targets, hit or miss will be signalled after each shot; in practices in which two shots are fired at one run, after every two shots.

500. *Method of signalling for iron bull's-eye targets.*

Signal.	Method of signalling.
Bull's-eye ...	Black disc placed on the dummy, or white disc on the target.
Inner	White disc placed on the dummy, or black disc on the target.
Outer	White disc, shown first at the left of the dummy, twirled round, and then placed on the dummy; or black disc used in the same manner and placed on the target.
Ricochet	Red flag, waved in front of target.
Miss	Red and white flag shown on the same side as the direction of the miss. If the direction cannot be determined, the flag is to be waved across the face of the dummy or target.

Hits will be valued as in para. **495**.

501. When telephones or mirrors are provided, all officers and non-commissioned officers will be instructed in their use, under battalion arrangements.

502. Where telephones are not provided, communication can best be maintained by means of a system of reflecting mirrors (*see* para. **491**), which will enable the officer in the butts to observe signals, made according to a pre-arranged code, from the firing point. Where neither telephones nor mirrors are available, bugle sounds must be utilized.

CHAPTER VII.

FIELD PRACTICES.

Section 93.—General.

503. In range practices, the soldier should attain a high standard of skill in shooting at known distances, under easy conditions and in various positions, at large vertical targets, easy to see, and furnished with scoring or approximation rings which enable the error in shooting to be expressed in figures, convenient for comparative purposes.

He should have confirmed in practice the lessons learned in preliminary training, and be thoroughly acquainted with the peculiarities of his rifle.

He has fired in the open and from behind cover, in slow and rapid practices, and should have learned the rate of fire which in his own case best combines volume with accuracy. In snapshooting he has been brought to realize the necessity for rapid alignment of sights, and the value of time in taking advantage of targets exposed under service conditions.

504. At this stage of training there appears very often, among those who have developed a degree of skill above the ordinary, a tendency to specialize in the elementary form of slow shooting, and to regard as the ultimate object of musketry training what is, in fact, only a means to an end. Deliberate practice at an elementary target tends to inculcate a slow method of shooting, as minute attention to changes of wind and light are necessary to produce the best results under such conditions. Although experts in this form of shooting gain an admirable control of their nerves under rifle range conditions, they come to regard active exercise and timed firing as antagonistic to nerve control, and owing to want of practical training frequently fail under more difficult conditions to produce as good results as 2nd class shots. High scores in range practices bear no relation whatever to the results to be expected when firing under service conditions, even in peace time. The fine adjustment of the backsight is based on careful signalling of the position of every shot fired ; but in collective firing there is little opportunity of ascertaining the result of any one shot, and at medium ranges correction of sighting depends upon observation of the nucleus of the cone of fire or of the effect produced by fire on the enemy's movement.

505. False standards of fire effect set up under artificial conditions lead to misconception as to the value of long range fire against low service targets, and individual firing may, in consequence, be employed where effective results can only be obtained by a liberal expenditure of ammunition, or time and ammunition may be wasted in useless firing at long ranges. The probability of error in judging distance and atmospheric influences at distances beyond 800 yards is such that there is little to be gained by developing a high standard of accuracy in shooting at such ranges. The difficulty in observing results is so great that a satisfactory degree of assurance of fire effect can only be obtained, beyond (about) 600 yards, by means of collective fire.

506. Fine judgment of wind and light as acquired in deliberate shooting, and minute adjustment of the sights based on shot for shot marking, will return less value in war for the amount of energy expended in peace training than the study of fire direction in all its aspects, of reconnaissance, ranging, collective grouping, and observation of the cone of fire. Range practices are, therefore, in no sense a final training; it is essential that further practice should take place under service conditions.

507. In view of the importance of executing field practices at unknown ranges, general officers commanding-in-chief will, when classification ranges only are available, endeavour to hire suitable ground for these exercises, under the Regulations for Engineer Services, Peace (Part I).

508. Every care should be taken to develop further the skill already acquired in snapshooting and rapid firing. The nerve control gained in deliberate shooting should be supplemented by vigour and alertness, deftness in loading, the habit of correct action under distracting conditions, and skilful use of cover, based on the determination to make fire effective.

509. There must be further training in picking up an indistinct target, such as is likely to be presented in war, in estimating its range, in rapidly opening fire, and in making the best use of ground; every individual must learn to recognize the distances at which individual fire will be effective, and to act in co-operation with his comrades.

510. Officers and fire unit commanders must be practised in their duties of direction, control, and observation of fire, in the use of ground, and in mutual support. With these must be combined the study of the results to be obtained from the delivery of concentrated fire at targets representing troops in different formations, and on ground of varying character, in order that practical experience may be acquired of the principles which govern the employment of fire in the field.

511. Under the conditions of peace manœuvres with blank ammunition, fire control may be neglected, targets may be insufficiently described, and it is not known whether the firers recognize them, distances are sometimes roughly guessed, sights are not always adjusted, and men aim carelessly. Unless, therefore, tactical exercises are conducted sometimes with ball ammunition, there will be a want of realism in training during peace time.

Section 94.—*Special instructions.*

512. Not more than 20 rounds should be fired in field practices in one day.

513. The dress for individual field practices will be drill order, and for collective field practices marching order.

514. The firing positions will be any that are suited to the ground or conditions of the scheme.

515. Some practice should be given in snapshooting in the standing position, during rapid advances at short range.

516. The general programme of field practices should be arranged as follows:—

i. Individual field practices.
ii. Fire direction practices.
iii. Collective field practices, divided into:—
   (1) Exercises for platoons and sections in fire direction and application of collective fire.
   (2) Standard tests of collective grouping and fire effect.
   (3) Comparative demonstrations of fire effect and vulnerability.

(4) Exercises for companies designed to reproduce service conditions as far as possible, and to illustrate tactical principles.

iv. Combined field firing.

517. If it is absolutely necessary to use classification ranges, the practices should be performed on a flank, if the extent of the danger area will permit. In this case, or if the actual range is used, the ground should be broken by means of screens, earth parapets, brushwood, or any other suitable contrivance, with a view to introducing some realism and uncertainty as to distances.

518. The allotments of ammunition to individual and collective field practices respectively may be varied as general officers commanding may decide, but the total amount allotted to field practices is to be fired in field practices, even if a classification range only is available. Due safety precautions must be taken. (*See* M.R., Part II.)

519. Company commanders will keep registers of all collective practices fired; records of permanent interest will be kept in a book for information in subsequent years. Standard figures will be published in battalion orders from time to time as a guide to the number of hits to be expected under various conditions, and a permanent record of such information will be preserved in every battalion.

520. The company diaries and battalion records of standard and comparative firing will be examined by the commanding officer, and produced at the inspections of general officers commanding (*see* para. **629**). Ricochets will not be included in the figures recording results of standard or comparative firing. They will be shown separately from other hits in the registers.

521. The principles defined in this chapter will be found generally applicable for the training of small units of cavalry. But in the case of squadrons and larger bodies, where ground of sufficient dimensions is available, schemes should be framed to bring into prominence the circumstances under which recourse should be had to fire action, as laid down in "Cavalry Training."

522. The scope of the exercises will necessarily vary with the extent of ground at the disposal of the troops, but due consideration should be given to the practising of commanders in rapid occupation of successive positions from which sudden bursts of fire from an unexpected quarter can be brought to bear on the objective selected.

523. Special attention should be devoted to skill in making fire effective from the first opening, preliminary orientation and ranging, anticipatory orders for fire direction, rapid advance to the fire position, simultaneous opening of fire in full volume, and control of fire to ensure a means of rapidly breaking off an action if necessary.

Section 95.—*Individual field practices.*

524. In the individual practices, each firer will be provided with a separate target; he will learn to fire at unknown distances, depending on the observation of a comrade for information as to the result of his shots; he will fire at targets representing an advancing enemy, and will advance himself, firing at each halt. He will learn to use ground for fire effect and cover, to pass all orders and information received, to recognize the limits for effective individual fire and the principles which govern the choice of targets in individual fire.

525. Practice will be afforded in acting promptly against targets appearing suddenly and disappearing after a short period of exposure.

526. The value of every shot will be ascertained by markers, and notified to the firer. The men will, as a rule, fire alternately, working in pairs for mutual assistance. When a successful shot is observed, the correct sighting as found will be immediately notified by the firer to the remainder of the squad.

527. It is essential to give full information to the firers as to the object of the practice, and to criticize freely the good and bad points of their performance.

528. Any preliminary information as to the ground and results of range finding at long range, which might be available on service, should be given in the later practices in order to combine all methods of ranging.

529. The distances should not exceed 600 yards. Training should be progressive in regard to targets, distances, and all other respects.

530. The firers should be formed into small squads, but there should be no fire control, nor any orders except such as may be necessary to regulate fire and movement in the interest of safety and to ensure that the objects of the training are fully carried out.

531. It will be well to conclude the individual field practices with a demonstration of the comparative inefficiency of individual fire at distances beyond 600 yards. This may be effected by detailing individuals to fire at low service targets and noting the time required to produce any required effect, and then applying collective fire at the same targets under proper direction and control of fire leaders, equipped with field glasses.

Section 96.—Fire direction practices.

532. When the skill and determination of his men are such as to give a fire leader a reliable distribution of fire upon which to base his fire direction, the value of the fire will depend almost entirely on his ability to apply the fire to the target. If he makes an error in sighting equal to one-half of the depth of the effective zone, the target will receive little injury; if he fails to make the aiming point clear, there may be no opportunity for observing results, or such results as are observed may be misleading.

533. It has been found that under certain conditions not uncommon in war, as many as 75 per cent. of rounds fired at ranges beyond 400 yards are aimed at the wrong target or no target at all. This loss of fire effect may be almost entirely avoided by skilful use of field glasses, reconnaissance, reservation of fire, and clear description of aiming points; it is evident, therefore, that fire direction and discipline imply much more than correct sighting, and steady shooting—there must also be mutual understanding between officers, non-commissioned officers and men.

534. As the firers have graduated in preliminary training, in range practices, and in individual field practices, so the fire leaders must make a special study of fire direction before undertaking the direction of collective field practices.

535. Theoretical study alone will not be sufficient, but it is unnecessary to expend large quantities of ammunition in order to gain practical experience of the difficulties of ranging, judging atmospheric influence, describing and recognizing aiming points, observation of fire, and verifying sighting by trial shots, all of which combine to form an important part of the subject of fire direction at effective and long ranges. About 300 rounds should be sufficient.

536. Fire direction practices should be carried out by parties of officers and non-commissioned officers when ground is available. It is not necessary that the area over which firing takes place should ordinarily be used as a range, but special precautions must be observed if a manœuvre area is utilized and an ample danger area must be temporarily secured and closed to traffic.

537. Even if ground for firing at long ranges is not available, fire direction exercises should be carried out without firing, before commencing collective field practices.

538. Firing should begin at unknown distance, and the target should consist of a screen or any visible object, the results being observed and signalled by special observation parties. Range finders should be freely used, and the sighting, when verified, compared with the range as found by the range finder and the correct range as measured on a map. Lines of fire should vary, if possible, so that additional practice may be obtained in judging atmospheric influence. Every officer and non-commissioned officer should write down his orders for fire direction before firing takes place. The practices should be conducted in strict accordance with the tactical and strategical principles laid down in the Training Manuals.

539. At the conclusion of the fire direction practices, all fire unit commanders should be familiar with the effects of winds and temperature in shooting at 500, 1,000, 1,500, and 2,000 yards. The range tables should be studied, and lectures given on the influence of ground, ranging errors, and other details of fire direction.

Section 97.—Collective field practices.

540. In preparing a programme for collective field practices regard must be paid to the amount of ammunition available. The greatest amount of instruction will be obtained for the ammunition expended if a battalion programme is framed and a series of tactical demonstrations is carried out by selected units in the presence of the remaining units, but it is essential that all fire leaders should be exercised in the direction and control of fire, and that every man should be practised in the delivery of controlled fire while using ground for concealment and making short advances at top speed.

541. Any omission or neglect in fire direction, as well as any failure to adjust the sights should, at this stage of training, be regarded very seriously.

542. Collective field practices are primarily intended to afford the commanders of fire units practice in their duties of direction and control of fire. With this must be combined the study of the results to be obtained from the delivery of fire at targets representing troops in different formations, on ground of varying character, in order that practical experience may be acquired of the principles which govern the employment of fire in the field.

543. Individual field practices are conducted at distances less than 600 yards; collective field practices, if ground is available, should be fired almost entirely at longer ranges. When ground is not available for firing at ranges beyond 600 yards, practice in collective firing is necessarily conducted at shorter ranges, but such practice must be supplemented by fire direction practices without ammunition at longer ranges.

544. The principal points to be considered are the choice of targets, the justification or necessity for opening fire, the volume of fire required to effect the object in view, the methods of ranging, the orders for fire direction, timing of movement, mutual support, regulation of volume of fire, concentration or distribution of fire, necessity for searching or distribution in depth, the probable dispersion of the cone of fire, probable error in ranging and judging atmospheric influence, the description and recognition of target and aiming point, reinforcement, ammunition supply, mutual assistance, passing of orders and information and communication with the flanks and rear.

545. Reflectors are of great use in enabling men to watch the advance of their comrades from the target position ; they are thus enabled to detect errors in use of ground, to appreciate the value of sustained fire for mastering the fire of the defence as well as the advantages of brief exposure for avoiding aimed shots.

546. The practice of requiring units to arrange targets or positions for others to fire at, or during a skirmish to place head and shoulder targets on the ground to represent themselves in position at any temporary halt, has been found to stimulate interest.

547. Falling or collapsible targets are of great value in all field practices, and may be used with advantage in casualty competitions designed to test the relative abilities of two firing lines, which simultaneously fire at separate sets of targets representing their opponents. Each man is represented by a target placed in front of the opposing firing line and becomes a casualty if that target falls. In this way superiority of fire is soon established by one line or the other, and fire ceases.*

Section 98.—*Tactical schemes. Section and half-company exercises.*

548. The earlier exercises for the smaller fire units will be carried out under simple tactical schemes framed by the company commander, and arranged so as to give a progressive training to all ranks engaged. Schemes will be so drawn up as to give separate instruction in each phase of the combat, rather than to combine in each exercise all the operations included in the execution of a successful attack. They will also be designed to illustrate the various situations which may be expected on active service. They should test the proficiency of leaders in making fire effective on first opening and in regulating the volume of fire in accordance with the situation.

549. Disappearing targets should be used for defence practices. Targets for attack should be disappearing and collapsible. Orders should be sparingly given, so that all ranks may be accustomed to think for themselves. It should be sufficient as a rule if the superintending officer, after full explanation of the scheme, limits his instructions to the direction of the advance, limits of fire, first opening of fire, and general regulation of movement and reinforcement.

550. All ranks should consider themselves as being under fire when within sight or effective range of the targets, and all criticism should be suspended once the practice has commenced until the results are made known on conclusion.

* For further instruction as to the choice and arrangement of targets for field firing *see* Part II of these Regulations.

551. Mutual support and combined action should be frequently practised, with and without the exercise of control.

552. The scheme, as a rule, will illustrate either attack or defence. If the firers are to be in movement as in attack, rehearsal will be desirable, or if stationary, and the targets are to represent an enemy advancing or retiring, there should be preliminary working of the targets.

553. The firers may be most easily interested by drawing up a practical scheme, explaining it thoroughly beforehand, and notifying afterwards the results that should have been attained according to peace standards.

554. If the earlier collective field practices show that officers and non-commissioned officers are sufficiently trained in directing and controlling fire under service conditions, as far as they can be simulated in peace time, a quantity of ammunition should remain available for expenditure in standard tests of fire effect, as well as comparative demonstrations of fire effect and the vulnerability of different objectives.

Section 99.—*Standard tests of fire effect.*

555. A few rounds should be fired at various substances used for protective purposes, such as trees, stones, walls, and sandbags, to test penetration. Object lessons by selected men in use of ground for fire effect, in rapid adjustment of sights and change of objective, and in rapid but accurate fire, will produce a deeper impression on young soldiers than theoretical instruction with blank ammunition.

556. One company at least may be tested with advantage in grouping collective fire on screens arranged in three or more rows, spaced to avoid double hits.

557. Most favourable conditions should be allowed, and sighting should be verified by trial shots in order that any error in fire direction may be obviated. This demonstration of grouping should be used to emphasize the responsibilities attaching to fire unit commanders and the rank and file respectively in regard to the application of fire by the former and the grouping of fire by the latter. Further deductions as to grouping may be drawn from demonstrations carried out under more practical conditions, provided that a sufficient number of targets is used to measure the dispersion of the cone of fire.

558. The number of hits to be expected on any targets may be calculated by comparing the target surface with a diagram showing the normal grouping of collective fire on a vertical target, allowance being made for the distance of the nucleus from the centre of the diagram.

559. It is, however, best to compile in every battalion a record of normal and best results obtained by firing at various targets and formations, so that figures for comparison may always be available.

Section 100.—*Comparative tests of fire effect and vulnerability.*

560. Though peace results give no reliable indication as to probable war results, they may afford a sound basis for useful deductions as to the relative values of different forms of fire and the vulnerability of various formations, when the figures of one practice are compared with those of another practice conducted under similar conditions.

561. Thus there may be tests of the relative vulnerability for extended and close formations, of the different firing positions and forms of cover, of figures exposed to aimed fire during short rushes or longer rushes and prolonged advances and troops halted in exposed fire positions; the penalties attaching to wrong use of ground may be demonstrated in comparison with more skilful movement; the relative values of cover from view and cover from fire, the immunity from fire gained by making dummy trenches and false loopholes and invisibility produced by artificial screening of earthworks may be practically demonstrated. It should be clearly shown that rapid movement in extended order may often afford a better means of crossing a fire-swept zone than frequent halts and struggles for superiority of fire.

562. Useful information regarding fire effect will be gained by comparative experiments with slow and rapid fire and intermediate rates; by testing also the effectiveness of extended lines relying on concealment and use of ground, in comparison with dense firing lines trusting to volume of fire. There may be demonstrations of loss of effect due to change of objective or aiming point, of the assurance of effect gained by the use of combined sights, of the neutralising effect gained by distribution of fire, of the decisive effect obtained by concentration, of the enhanced results produced with oblique and enfilade fire, of the possibilities of indirect and night firing, of the special arrangements for observing and signalling results, of the use of long range sights, of the effect on the dangerous zone of inclination of ground with reference to the line of sight, and similar studies in fire tactics.

563. At the conclusion of a collective field practice, in addition to criticism of the conduct of the exercise, there should be a conference as to the conclusions to be drawn from the results. For this reason a complete record must be made of the targets and conditions; the figures showing results must be carefully and fully tabulated.

564. The criticism of the conduct of the exercise should be complete and should deal with the application of the tactical principles laid down in the Training Manuals as well as with the application of the principles contained in these regulations. In appreciating results, chief attention should be paid to the successful or unsuccessful result of the first application of fire, since surprise effect is all-important and correction of sighting by observation is rarely possible in war.

565. The percentage of hits to rounds fired is an index to the steadiness of the firing only if the fire direction has been proved to be satisfactory; if the fire direction fails, the more accurate the shooting the fewer will be the hits recorded.

536. In considering the results of fire, the percentage of loss inflicted on the enemy within a limited period of time is the best means of judging the value of the fire.

567. For general comparison of the collective fire results of units, the average number of hits per man per minute should be calculated if fire was concentrated, or the average number of figures hit per man per minute if it was distributed, but due regard must be paid to the justification of the rate of fire as indicated by the scheme.

Section 101.—*Company exercises. Combined field firing.*

538. When all the fire unit commanders have shown proficiency in fire direction and control, companies will be trained to fire collectively against firing lines at effective ranges. The leaders will be exercised in judging distance, in describing targets, and in concentration or distribution of fire; great weight will be attached to the accurate passing of orders and information and to the quickness of the men in recognizing their targets and applying fire.

Orders for fire direction and control should anticipate events as far as possible, so that the sights may be adjusted in readiness for any movement or favourable target.

The advance under service conditions should commence at distant range.

Range-finders will be freely used at long ranges for preliminary reconnaissance, and ranges noted for guidance in the further advance.

569. Combined field firing carried out under favourable conditions of ground by a force not larger than a brigade of infantry, a brigade of field artillery and a cavalry regiment,* teaches valuable lessons in co-operation between the various arms. It enables artillery commanders to study the adaptation of artillery fire tactics to the requirements of the other arms under realistic conditions and the other arms to become acquainted with the degree of support which can be given them by an efficient artillery.

570. The issue of elaborate safety regulations tends to tie too much the hands of the commanders responsible for carrying out these exercises and should be avoided. The directing officer should set a tactical scheme on which the commander of the troops taking part in the exercise should base and issue his orders as on service. The execution of these orders should be regulated by the umpires only when they consider that firing is or is likely to become dangerous to the troops or to the public.

---

* At stations abroad, where exceptional facilities as to ground exist, the size of the force may be increased at the discretion of commander.

571. General officers commanding-in-chief are authorized to expend 15 per cent. of the annual allowance of artillery ammunition for combined field firing. The small arm ammunition for the use of cavalry and infantry in combined field firing, will, at the discretion of the general officer commanding-in-chief, be drawn from those rounds allotted in Table B for collective field practices and surplus, and from those authorized in Equipment Regulations, Part I, para. 202. The expenditure is to be limited to a maximum of 15 rounds per man.

CHAPTER VIII.

COMPETITIONS, BADGES, PRIZES & METAL FUNDS.

Section 102.—*Competitions.*

572. All rifle meetings or competitions, towards which assistance is given directly, or indirectly from public funds, will be conducted in accordance with the spirit and training methods of these regulations.

573. To fulfil its object a rifle meeting should be so organized as to mould and develop military opinion in regard to the methods of rifle fire best suited to modern requirements ; it should at the same time encourage all ranks to improve themselves in the performance of those duties which will devolve upon them in war.

574. The attainment of a good average standard of proficiency by a large number of men is infinitely more important than phenomenal skill developed by a few champions, or the compilation of scores in excess of any previously recorded by competitors who confine their practice to one form of shooting.

575. The object will not be attained unless the conditions of each competition are suited to the rank and experience of the competitors.

576. Championships lend interest to a meeting, but do not directly assist the object in view. If a championship is awarded to an experienced competitor for skill in applying fire to a bull's-eye target when the range is known and each shot is signalled, money and ammunition are wasted, young soldiers are discouraged, and false ideas as to musketry training are disseminated.

577. When drawing up a programme the elements as well as the perfection of skill with the rifle should be kept in view, in order that elementary competitions may be designed for the younger competitors, while championships are awarded for perfected skill only.

578. The classification of competitors is most important: it should have regard to (*a*) duties in war; (*b*) experience. It is by the standard of skill in shooting, displayed by the younger competitors as a whole, and the proficiency of fire leaders in fire direction, that the success of a rifle meeting should be judged.

579. Individual competitions for the rank and file should include tests of skill in applying the various kinds of fire under conditions approaching those which would confront the individual soldier on active service.

580. Competitions for individual officers and section commanders should include only tests of skill in those forms of ranging and shooting which lead directly up to proficiency in fire direction. They may include sighting competitions at distances beyond 500 yards, observation of fire, appreciation of fire limits, shooting at unknown distances, range finding, and judging distance.

581. Since the direction and control of massed fire is the special duty of officers and non-commissioned officers, these requirements should be held in view in the preparation of team competitions, which should take place under conditions of advanced training.

532. Team competitions at unknown ranges are the most practical of all forms of shooting, but may lead to waste of ammunition, unless mistakes in fire direction are noted and penalised before fire is opened. If the error in sighting, as ordered, exceeds 100 yards, and the ground is unfavourable to observation of fire, it will be best for an umpire to correct the elevation and deduct points as may be arranged.

533. Individual competitions at long range have little military value; but when conducted as sighting competitions at several ranges, fire being limited to about three shots at each range, they afford a satisfactory trial of the ability of officers and non-commissioned officers to estimate the influence of the atmosphere, and call for thorough knowledge of the rifle.

534. Bull's-eye targets should, as a rule, be used in elementary competitions only for young soldiers in the first year of their service. They may, however, be used in competitions similar to those suggested in para. **533**, or in match rifle competitions designed to test the powers of rifles and ammunition, rather than the military skill of the firer.

535. Considerations of scoring should not be allowed to influence unduly the design of targets; service visibility is the first consideration; figures or bands are preferable to discs; scoring rings should be invisible, and neutral colours should be adopted.

536. Programmes should, as far as possible, embrace all the elements of musketry training. The disproportionate value sometimes attached to proficiency in grouping and applying fire under easy conditions, leads to neglect of further training.

587. Sighting shots should only be allowed in those rapid firing competitions in which accuracy combined with a high rate of fire is the first consideration.

588. Specializing in any one kind of shooting should be discouraged; competitions for trained soldiers should include not less than three different forms of shooting, such as snap-shooting, sighting, attack, rapid firing, observation, appreciation of fire limits, &c., with as much variation of targets and conditions as can be conveniently arranged.

589. It is often impossible to hold competitions at an unknown range owing to want of range accommodation. In such a case it is very desirable that a special competition in judging distance be held in connection with, but not necessarily on the same days as, the shooting competitions, and every means should be adopted of emphasizing its importance.

590. The value of steady, accurate shooting should be kept in view as the foundation of skill with the rifle, but the paramount importance of time in military shooting must not be overlooked in formulating the conditions for a championship.

591. Conditions for soldiers with more than one year of service should never permit a slower rate than three rounds a minute for a series of shots, exclusive of time taken for signalling. In slow shooting the best military shot is probably the man who, in skirmishing, first applies an effective shot to a service target at unknown range; in rapid firing the sighting to be used would very often be communicated to the soldier on service, and a high rate of fire, combined with reasonable accuracy, would be expected from him.

592. Ranges will not be allotted for practice under elementary conditions (except for young soldiers) if required for shooting of a more practical nature.

593. Coaching should never be allowed, but individual soldiers in observation and skirmishing competitions may be allowed to work in pairs, giving mutual assistance.

594. Rules should never be framed to exclude difficulties such as are met with on service, with the object of eliminating elements of chance. Such a course will remove what is of most value—the personal test of skill in meeting service difficulties. An element of chance is accepted and regarded as advantageous in competitions of a sporting nature, and will increase interest in military competitions, since it is always present in war.

595. It is essential that in competitions open to the rank and file, competitors should be allowed to fire only with a rifle on charge of their unit, and no departure from the regulations governing the painting of sights, use of slings, provision of wind flags, targets, rifle accessories, &c., is to be permitted in service rifle competitions.

Section 103.—*Badges.*

596. A marksman's badge of crossed rifles will be issued to marksmen of all branches of the Regular Army, Special Reserve, and Territorial Force, to the infantry of the Channel Islands Militia, and to the King's Own Malta Regiment of Militia, provided such marksmen are below the rank of warrant officer. This badge can only be awarded for performances in Part III., Table B, Appendix I., or its equivalent.

597. Good shooting badges, for which all ranks except officers and warrant officers are eligible, will be issued to cavalry, Royal Engineers (Sappers) and infantry of the Regular Army; to Irish Horse, King Edward's Horse and infantry of the Special Reserve; to the infantry of the Channel Islands Militia, to the King's Own Malta Regiment of Militia, and to Yeomanry, Royal Engineers (field companies) and infantry of the Territorial Force, under the following conditions:—

i. *Best shot in each squadron or company.*—A badge of crossed rifles and star. Issuable to the non-commissioned officer or man in the squadron or company (including band) who has made the best score in Part III., Table "B," Regular Forces, in which company or squadron at least 30 men have been exercised in Part III., Table "B," during the musketry year. Casuals or men attached from other units or companies are not eligible.

ii. *Best shot of the serjeants and lance-serjeants in the regiment or battalion.*—A badge of crossed rifles and crown, surrounded by a scroll of bay leaves. Issuable to the best shot of the serjeants and lance-serjeants in any regiment or battalion in which all squadrons or companies have exercised at least 30 men in Part III., Table "B," Regular Forces, during the musketry year.

iii. *Best shot of the corporals and privates in the regiment or battalion.*—A badge of crossed rifles and star, surrounded by a scroll of bay leaves. Issuable to the best shot of the corporals and privates in the regiment or battalion in which all squadrons or companies have exercised at least 30 men in Part III., Table "B," during the musketry year.

iv. *Section commanders of the best shooting company.*—A badge of crossed rifles and crown. This will be awarded to the section commanders of the best shooting squadron or company in each regiment of cavalry or battalion of the Royal Engineers or infantry in which all the squadrons or companies have exercised at least 30 men in Part III., Table "B," Regular Forces, during the musketry year.

For a N.C.O. to win this badge, he must have been exercised in Part III., Table "B," during the musketry year with his squadron or company.

In the case of *Royal Engineers.*—(1.) Regulars—

The "battalion" referred to in sub-paras. ii, iii and iv* is constituted as follows:—

(*a*) Depôt Battalion, Chatham.
(*b*) Training Battalion, Chatham.
(*c*) Signal units quartered in Aldershot Command.
(*d*) Field Companies quartered in Aldershot Command.
(*e*) Field Companies quartered in Ireland.
(*f*) Survey Companies and "K" Signal Company.
(*g*) Fortress units in South-Western Coast Defences, Western Command, and Channel Islands.
(*h*) Fortress units in Southern Coast Defences.
(*i*) R.E. units quartered in each Command, excluding those specified above, where not less than three units have been exercised.

Units quartered—

* In the case of Royal Engineers the badges mentioned in sub-para. iv are issuable to section commanders of the best shooting company of the battalion.

(*j*) In South Africa and Mauritius.
(*k*) At Gibraltar.
(*l*) At Malta.
(*m*) At Hong Kong, Singapore and Ceylon.
(*n*) At Bermuda, Sierra Leone, Jamaica and Egypt.

The following officers will decide which is the best shooting company in each battalion, viz. :—

(*a*) and (*b*) The Commandant, S.M.E.
(*c*), (*d*), (*e*), (*h*), (*i*), (*k*), (*l*) The local Chief Engineer.
(*f*) The Director-General, Ordnance Survey.
(*g*) The Chief Engineer, Plymouth.
(*j*) The Chief Engineer, South Africa.
(*m*) The Chief Engineer, South China.
(*n*) The Chief Engineer, Egypt.

(2.) *Territorial Force* (*field companies*).—A "Battalion" will consist of the field companies in any one Command.

598. *Machine-gun badges.*—Regular Cavalry and Infantry. A badge for first-class machine gunners consisting of the letters "M.G." within a wreath to be worn on the left forearm.

The machine-gun badge may be worn in addition to and above the marksman's badge, but may not be worn in addition to the badges mentioned in para. **597**, i, ii., and iii.

599. The selection of the best shooting squadron or company should not depend only on the results of the range and field practices, but when possible, the commanding officer will be guided in his selection by the results of his own personal observation.

600. These badges will be worn as follows:—That mentioned in para, **597** iv., on the right forearm; the marksman's badge and those referred to in para. **597** i., ii., and iii., on the left forearm, the one which is obtained for the highest distinction being nearest the cuff. The marksman's badge will not be worn in addition to the badges mentioned in para. **597** i., ii., and iii., but the badge awarded to the section commanders of the best shooting company will be worn in addition to any other badge.

601. The names of all entitled to good shooting badges will be published in regimental orders.

602. Badges for good shooting will be issued as soon as possible after they have been won.

603. Ties will be decided by the commanding officer.

Section 104.—*Judging distance badges.*

604. A badge for proficiency in judging distance, consisting of a star to be worn on the right forearm above any other badge, will be issued as follows:—

i. To the four best privates in each squadron of cavalry, to the three best in each company of Royal Engineers and to the six best in each company of infantry, as demonstrated by their performances in the four quarterly tests referred to in para. **323** (not less than 16 distances).

ii. To the four best non-commissioned officers in each squadron of cavalry, to the three best in each company of Royal Engineers and to the six best in each company of infantry, selected by the unit commander according to the results of the quarterly tests; not less than 16 distances between 200 and 1,400 yards will be judged.

iii. The award will be published in orders after the fourth

test in each year, and the badges will be at once issued and worn until the next award.

(This section will not apply to depôt squadrons or companies.)

Section 105.—*Prizes.*

605. The grants allowed by the Royal Warrant for Pay, &c., for issue as musketry prizes to NON-EUROPEAN ENGINEERS AND INFANTRY will be awarded as follows:—

i. Grant (*a*) will be drawn for all recruits (except officers) who complete Table "A," and will be awarded to the best shots of each party as the commanding officer may deem desirable.

ii. Grant (*b*) will be drawn for all ranks (except officers and warrant officers) who have completed the classification practices of Table "B." It will be awarded in accordance with a scheme which will be prepared by the commanding officer at the commencement of the musketry year, and submitted for the approval of the general officer commanding.

iii. All those in respect of whom money is drawn will be eligible to receive prizes, but serjeants should compete separately, and not with the rank and file.

iv. As soon as a company has completed the classification practices, Army Form O 1716 will be prepared, in triplicate, and forwarded for the approval of the general officer commanding. When approved, two copies will be returned to the unit, one for retention and the other for transmission with the pay list in which the prizes are charged against the public.

v. Similar procedure will be followed in regard to parties of recruits.

606. The awards will be notified in orders.

607. In regard to the forfeiture of prizes, *see* Royal Warrant for Pay, &c.

608. The general officer commanding is authorized to decide in respect to the issue of prizes in cases in which he may consider their issue unmerited.

Section 106.—*Metal funds.*

609. A fund formed from the sale of metal recovered from rifle ranges will be established in each command.

610. The metal will be disposed of to the best advantage under the orders of the general officer commanding, due regard being paid to the Woolwich half-yearly price list for old metal. In the case of stations abroad the metal may under certain circumstances be returned to Woolwich for disposal (*see* Regulations for Army Ordnance Services, Part I).

611. The collection of metal from the butts is the duty of range wardens, who should, in order to ensure that none is lost, pick the stop butts daily, immediately after the cessation of fire, and lodge the metal in a place of safety. Troops will only be employed in fatigue duty in this connection when it becomes necessary to reface the butts.

612. In the case of ranges in barracks, metal will be collected under regimental arrangements, and disposed of as may be directed by the general officer commanding.

613. After payment of any necessary expenses connected

with the prevention of theft of metal, &c., the sums realized will be allotted by general officers commanding for such purposes as prizes and payment of markers at rifle meetings.

614. Should the sums accruing from the sales exceed the foregoing demands, general officers commanding are authorized to expend any surplus on the provision of sub-target rifles, the preparation of ground for field practices, field firing, and other aids to musketry training for which funds are not otherwise provided, but not on the purchase of ammunition; the money so expended will be shown on Army Form N 1472.

615. Army Form N 1472 will be completed at the end of each financial year and rendered to the Local Auditor attached to the command, or, in the case of commands in which there is no Local Auditor, to the Assistant Financial Secretary, War Office.

CHAPTER IX.

REPORTS, RETURNS AND RECORDS.

Section 107.—*General.*

616. The following returns are to be used for recording musketry performances. Records and diaries of collective field practices and Army Books 90, 100, 107 and 218, will be retained until out of date. The Regimental Company and Recruits' Annual Returns and records of all tests should be retained for three years; general officers commanding may authorize the destruction of all others, including Army Form B 186, after the conclusion of the musketry year.

Section 108.—*Annual returns.*

617. *Recruits' musketry return.*

Army Form B 188 (*Cavalry, R.E., and Infantry* [*Regulars*], *and R.E., and Infantry* [*Special Reserve*]).

i. The names of the men will be entered on this return in order of squads on the day on which a party of recruits commences the practices of Table A. A blank line will be left between every two names, in which repetition scores will be entered. Scores will be entered daily.

ii. A line will be drawn across the return to separate one party from another.

iii. At the end of the musketry year (para. 8) the summary will be completed and the return signed.

The total points made by all the parties exercised in Part IV will be added together, and the average individual score found, for record in the regimental annual return.

iv. In the Special Reserve recruits will be classified on the conclusion of the practices of each party.

618. *Recruits' musketry and battery annual returns.*

Army Form B 192A (*R.A. and A.S.C.* [*Regulars and Special Reserve*], *and A.O.C.*).

The instructions given in para. **617**, modified where necessary, are to be followed in the preparation of this form.

619. *Company annual returns.*

Army Form B 192 (*Regulars and Special Reserve*).

i. The name of every officer, warrant officer, trained non-commissioned officer, and soldier borne on its strength will be entered on the day on which a company commences the range practices of Table B; officers first, other ranks by squads, and after these the names of all, including men exempt, who for any reason are not being trained with the company.

ii. On the conclusion of each day's range practice the scores obtained will be entered from the registers.

iii. When the "Qualifying Practices," Part I, have been completed, the scores of men who failed to obtain the qualifying points will be crossed through with a single fine line on pages 2 to 6 and their names re-entered in the space provided on page 1, and their scores on repetition will be there recorded.

iv. In Part II only the number of rounds fired will be entered against each man's name.

v. The performances of men who are transferred from one company to another in the same unit, or to another unit after the completion of Part III of Table B, will be retained, for record, in the return of the company by which they were exercised; those of men who have not completed this part will be ruled out of the return of the company giving, and entered in that of the company receiving, the transfer.

vi. The performances of casuals and the names (and points, if necessary) of men transferred to the company will be entered as soon as transfer returns are received.

vii. In the column for Parts IV and V, the number of rounds fired in each practice will be entered against the name of each man who takes part, and a concise description of each practice will be given on page 7. A description will also be given on page 7 of the fire direction practices carried out (Part V), stating the total number of rounds used in these practices.

viii. When the register is kept at the firing point, and a memorandum at the butts (para. 630), the necessary additions and deductions will be made before the company average is struck.

ix. Immediately after the end of the musketry year (para. 8) the return will be totalled and fully completed.

x. The blank columns against the names of partly-exercised and non-exercised men will be ruled through, and the reason noted in the column for remarks.

xi. Officers commanding companies will give full particulars of any practices fired in Part II which are not in accordance with those described in Table B.

xii. Averages are not to be worked out in qualifying or instructional practices.

xiii. Each man will be classified on the day he completes Part III, or, if for any reason he fails to complete Part III, then on the last day of the year.

620. *Regimental annual returns.*

Army Form B 187 (*Cavalry, R.E., and Infantry* [*Regulars*]), (*R.E., and Infantry* [*Special Reserve*]).

Army Form B 187A (*R.A., A.S.C.* [*Regulars and Special Reserve*], *and A.O.C.*).

i. These returns will be prepared from the company and recruits' returns, and be sent through the usual channel within 14 days of the conclusion of the musketry year, in triplicate, to the general officer commanding-in-chief, who will enter his remarks, send one copy to the commandant, School of Musketry, Hythe, return one to the unit, and retain one.

ii. When a unit is removed from one command to another at home after completion of its annual course of musketry (that is excluding casuals or small parties), the return will be submitted through the brigade and divisional commanders and the general officer commanding-in-chief under whom the course was carried out.

iii. The commanding officer will note in this return full particulars of any variations in the courses sanctioned under para. **439.**

iv. Detachments (and units not quartered in a station with other units of their corps) of the Royal Horse and Royal Field Artillery, Royal Garrison Artillery, Royal Engineers, Army Service Corps, and Army Ordnance Corps, will render returns on Army Form B 187 and B 187A. Where several companies are stationed together, one return may be rendered for a brigade or for such groups of units as general officers commanding may direct. Army Form B 187A will be used for Royal Artillery units of the Channel Islands Militia, for units of the Royal Army Medical Corps and Army Veterinary Corps in South Africa, and for recruits of the two latter corps at home.

v. General officers commanding will state in the return the arrangements made for independent supervision and marking during the practices of Part III.

vi. At home stations general officers commanding-in-chief will render to the commandant, School of Musketry, Hythe, for submission to the Army Council, by the 1st February in each year, separate reports, in narrative form, on the training of the Regular Troops, Special Reserve, and Territorial Force in their commands, together with the regimental annual returns.

vii. In these reports they will draw special attention to any circumstances which have debarred units from compliance with the regulations for effective training in musketry, and will mention units in which the training has not come up to the required standard, stating the directions they have given to ensure its attainment. Recommendations regarding matters which cannot be decided locally will be made on a separate sheet at the end of the report on each branch of the service.

viii. Reports and returns from India and other stations abroad will be submitted as soon as possible after the conclusion of the musketry year.

ix. Brigade commanders will submit, on these returns, reports on the training of individual units, based on inspection and on the results of firing conducted under classification and field practice conditions. The following points should be specially referred to:—

(*a*) The individual efficiency of the men as shown by their firing and independent action during the classification practices and individual field practices.

(*b*) The proficiency of officers and non-commissioned officers in judging distance and range taking, as well as their instructional capacity. Their general efficiency in musketry will be judged by the conduct and results of collective field practices, fired under conditions approximating to those of active service.

(*c*) The ability of the officers to give musketry training its full tactical significance, as indicated by their methods of instruction, lectures, essays, and practical demonstrations of fire power.

(*d*) The efficiency of the machine-gun sections, and general knowledge of machine-gun tactics.

621. *Musketry transfer return.*

Army Form B 193.

i. This form will be used in all cases of transfer, and to transmit the performances of casuals and of recruits to their company commanders; the registers recording the performance of a casual will be retained in the company exercising him.

ii. In the case of men who have not commenced Table A (or B) at the date of transfer, the words "recruit (or trained soldier), not exercised" will be inserted in the column for remarks, in which it will also be stated, in the case of a trained soldier, not exercised, whether he is qualified, in a musketry sense, for service or proficiency pay.

iii. In all other cases the return will show fully the practices performed and scores obtained.

622. ***Musketry prize return.*** (Non-European engineers and infantry of the Regular Forces and Militia.)

Army Form O 1716.

Instructions for the preparation of this return are given in para. **605**, and on the form.

Section 109.—***Registers.***

623. ***Register of judging distance tests.***

Army Form B 186.

This form will be used for the tests prescribed in paras. **323** and **324**. Instructions for its preparation are issued with the form.

624. ***Register of range takers' tests.***

Army Form B 66 or 66A.

This form will be used for the tests prescribed in Section 71.

625. ***Register for grouping practices.***

Army Form B 68.

This form will be carefully completed as a record of all grouping practices fired in Tables A and B.

626. *Company commander's roll for range practices.*

(For gallery rifle ranges. To be kept at the firing point.)

Army Form B 189.

i. This form will be used for all practices mentioned in paras. **477** and **478**.

Names will be entered in order of squads before proceeding to the rifle range, but the order of firing will not be inserted until men are detailed to shoot. Occasional shots will be recorded on the back. The names of men directed to cease firing under para. **418** will be re-entered at the end of their squad, and a fresh number in order of firing allotted to them.

ii. Should a practice be broken off, the company commander will rule out the names of those who have not fired, and fresh forms will be used when the practice is completed.

iii. The roll will be signed at the firing point.

627. *Butt register for range practices.*

(For gallery rifle ranges.) Army Form B 190.

i. This form will be used for all practices mentioned in paras. **477** and **478.**

ii. It will be kept by the officer on duty at the butts, in accordance with the instructions in para. **485.**

iii. On the conclusion of the practice a line will be ruled diagonally across the unused spaces, the form will be signed and handed to the company commander, who will attach it to his roll.

iv. The register will be the record of the scores obtained.

628. *Firing point register for range practices.*

Army Form B 190A.

This form will be used on non-gallery rifle ranges in the special cases referred to in para. **488,** and by the Territorial Force when it is not possible to arrange for the method under which the register is kept at the butts. In the case of 30-yards ranges no other record will be kept, except such as commanding officers and company commanders require for their own information.

629. *Register for collective field practices.*

Army Form B 62.

This register will be completed as soon as possible after return to barracks, and be produced by officers commanding companies for inspection when called for.

Extracts will be made as considered desirable by commanding officers for entry in battalion orders and in the battalion permanent record of collective field practices.

630. *Butt memorandum.*

Army Form B 67.

This form will be kept at the butts whenever the register is kept at the firing point. In the Territorial Force, when it is not possible to arrange for the presence of an officer or non-commissioned officer in the butts, its use will be dispensed with, and the register will be the only record of the scores obtained. In all other cases the memorandum will be the record of points obtained by the squad, the register of those obtained by individuals.

Section 110.—*Diaries.*

631. *Company commanders' diary of field practices.*

A complete record of all field practices fired will be kept with the registers by company commanders and will be inspected periodically by commanding officers.

632. i. *Company commanders' diary of ammunition expended.*

Army Book 99.

ii. *Assistant adjutant's diary of ammunition expended.*

Army Book 100.

For instructions, *see* para. **141.**

Section 111.—*Territorial Force returns.*

633. ***Adjutants' and squadron or company commanders' diary of ammunition expended.***

(Territorial Force.)

Army **Book 218.**

For instructions, *see* **para. 141.**

634. i. ***Recruits' musketry and squadron or company annual return.***

(Territorial Force.)

Army Form E 552.

ii. *Regimental annual musketry return.*

(Territorial Force and O.T.C.)

Army Form E 569.

Instructions for the preparation of these forms are printed upon them; they will be supplemented—with the necessary modifications—by the instructions respecting the preparation of Army Forms B 188, B 192 and B 187.

Army Form E 552 will be used for recruits and trained men of Royal Artillery, Royal Engineers (other than field companies) and Army Service Corps units.

635. ***Results of Territorial Force musketry training.***

Army Form E 567.

Full instructions for the preparation of this form are issued with it.

635A. *Contingent annual musketry return (O.T.C.).*

Army Form E 561.

This form is for use when the course is fired on the miniature range.

Section 112.—*Machine gun returns.*

i. Army Form B 2050.

636. This form is a register used at the firing point for machine gun practice.

ii. Army Form B 2050A.

This is a firing point register to be used in conjunction with Army Form B 2050.

iii. Army Book 107.

This book contains a record of the ammunition expended on machine gun practice.

CHAPTER X.

MACHINE GUNS.

Section 113.—*General.*

637. Details as to the mechanism of the gun are contained in the Handbook for the ·303-inch Maxim Gun. The drill of sections and machine gun tactics are dealt with in the Training Manuals, upon which the training of detachments will be based.

638. The depth of the zone beaten by 75 per cent. of shots fired from a Maxim gun, measured as described in para. **181**, may be taken as 150, 70, 60 and 50 yards respectively at 500, 1,000, 1,500 and 2,000 yards, and the lateral dispersion of the cone of fire at the same ranges as 4, 8, 13 and 19 feet.

638A. The following data will serve as guides in assisting machine gun commanders to make the necessary allowance for wind when giving fire orders :—

Mark VI Ammunition. *Right-angle winds.*

Range (Yards).	Allowances for—					
	Mild.		Fresh.		Strong.	
	Yards.	Feet.	Yards.	Feet.	Yards.	Feet.
500 ...	—	2	—	4	—	6
1,000 ...	3	—	6	—	9	—
1,500 ...	6	—	12	—	18	—
2,000 ...	12	—	24	—	36	—

Oblique winds.—Halve the allowance for right-angle winds beyond 500 yards range.

Head and rear winds.—Ignore within 1,000 yards; between 1,000 yards and 1,500 yards allow 50 yards.

Mark VII Ammunition.

When the wind is from the right the above data apply; when the wind is from the left halve the above allowances.

This is accounted for by the marked influence of drift with Mark VII ammunition.

In certain circumstances the above data may also be useful in directing infantry rifle fire to which they equally apply.

Range Table for Maxim Guns for Marks VI and VII Ammunition.

Range (Yards).	Angle of Tangent Elevation on the Gun.		
	Mark VI.		Mark VII.
	Gun Maxim ·303″.	Guns Maxim ·303″ converted Mks. I and II.	Guns Maxim ·303″ and converted Mks. I and II.
	° ′	° ′	° ′
100	10·5	10·0	12·5
200	14·5	15·0	15·0
300	21·0	20·5	18·5
400	27·0	26·5	22·5
500	35·0	34·0	27·0
600	44·0	43·0	32·5
700	56·5	55·0	38·5

Range (Yards).	Angle of Tangent Elevation on the Gun.		
	Mark VI.		Mark VII.
	Gun Maxim ·303″.	Guns Maxim ·303″ converted Mks. I and II.	Guns Maxim ·303″ and converted Mks. I and II.
	° ′	° ′	′
800	1 10·0	1 7·5	46·0
900	1 23·0	1 19·5	54·0
1,000	1 38·0	1 33·5	1 3·5
1,100	1 53·5	1 48·5	1 14·5
1,200	2 11·0	2 6·0	1 27·0
1,300	2 28·5	2 22·5	1 41·0
1,400	2 49·5	2 43·5	1 57·0
1,500	3 9·0	3 2·0	2 15·0
1,600	3 30·0	3 22·5	2 35·0
1,700	3 55·0	3 45·5	2 58·0
1,800	4 22·0	4 11·5	3 23·5
1,900	4 50·0	4 38·5	3 52·0
2,000	5 20·5	5 8·0	4 24·0
2,100	5 53·0	5 35·5	5 0·5
2,200	6 29·5	6 6·5	5 41·0
2,300	7 11·0	6 42·0	6 26·0
2,400	7 57·0	7 19·0	7 17·0
2,500	8 46·5	8 0·0	8 14·0
2,600	9 39·0		9 18·0
2,700	10 37·0		10 30·0
2,800	11 37·0		11 50·5
2,900	12 41·0		

NOTES.—(*a*) Heights of trajectory and angles of descent may be taken as being the same as for the S.M.L.E. Mark III rifle.

(*b*) Converted guns are not sighted for ranges beyond 2,500 yards.

639. Since the consistency of the ground, the condition of the barrel, and the play of the mounting affect in a greater or less degree the dispersion of the bullet cone, it is necessary that a firer should be well acquainted with the peculiarities of the gun and mounting.

640. In order to teach the use of the muzzle attachment (*see* Handbook for the Maxim Gun), 50 rounds should be fired annually with the attachment adjusted.

641. No man should begin firing with service ammunition until he has correctly passed the tests of elementary training.

It is important that all the points to be observed before, during, and after firing are carefully carried out, in order to render them habitual to all machine gunners.

The section officer will fire Part I, also Practices 7 and 8, as well as the classification practices of Part II. Non-commissioned officers will fire all practices of both Parts.

Tests of elementary training.

642. The following tests of elementary training have been devised to assist officers in testing the efficiency of their sections in elementary training, and also to ensure that no detail of such training is overlooked. It is important that these tests should not be considered as competitions against time, for although quickness is necessary, yet accuracy is the first essential. No man should therefore be passed as efficient unless all the points are properly fulfilled, even though he may complete them in the standard time.

The tests must be carried out in strict accordance with the detailed instructions given, for unless the smallest details are insisted upon, the time limit will not be applicable. In carrying out these tests, time can be saved if the first pair complete Tests i. to v. consecutively; the remainder can be carried out as convenient.

The entire personnel of a machine gun section, including, when possible, the drivers, should qualify in these tests, acting both as No. 1 and No. 2. This is necessary, for on service any member of a section may be required to replace a casualty at a moment's notice. In all tests, No. 1 will repeat all orders received.

i. *To erect the tripod and mount the gun on the command "Mount Gun."*

The tripod, gun, and ammunition box, to be laid on the ground with Nos. 1 and 2 standing one on each side of them. The clamps of the tripod legs to be sufficiently tight to prevent them from falling loose when lifted; they must be close enough together to enable the tripod to be put into the hood; the strap to be buckled round the rear leg; traversing clamp tight. The position where the gun is to be mounted to be not more than 5 yards away.

Points to be observed.

Crosshead to be upright; all clamps tight; pins home and turned down; both elevating screws exposed the same distance; gun pointing to the front; cork plug withdrawn; Nos. 1 and 2, and an ammunition box, to be in position; tripod adjusted so that the gun is at a suitable height for No. 1 to lay and fire in a comfortable position without constraint. Standard time, 20 seconds.

ii. *To load the gun on the command "Load."*

In continuation of i.:—Belt, with a few dummy rounds at the end, properly packed in the box, which will be closed and fastened.

Points to be observed.

All loading motions to be quite distinct and correct; to be carried out without any slurring. Standard time, 5 seconds.

iii. *To adjust the sights and lay the gun on the completion of the command* " *At (object)...... (Range).........*"

In continuation of ii.:—Gun loaded and ready to lay. Three objects will first be pointed out on a landscape target placed about 25 yards from the gun, but the No. 1 being tested will not know which will be given. Any range may be ordered so long as it will not be necessary to alter the slide by more than 500 yards up or down when the aiming mark is changed.

Points to be observed.

That the slide is adjusted and the gun laid with absolute accuracy. When checking the aim, " holding " pressure must be exerted on the handles; this may generally be done most conveniently by the instructor, but, should there be any question of different " holding," the No. 1 should hold the gun while the instructor checks the aim. No. 1 must be careful, however, that he does not exert lateral pressure when leaning to one side to clear the sights. Standard time, 12 seconds, from the range being ordered until No. 2 holds up his hand, indicating that No. 1 is ready to open fire.

iv. *To unload the gun on the command* " *Unload.*"

In continuation of iii.

Points to be observed.

Tangent sight lowered, but without moving the slide; unloading motions to be quite distinct, without slurring;

belt withdrawn, repacked correctly in the box with lid closed and fastened; lockspring released. Standard time, 5 seconds.

v. *To dismount the gun on the command "Dismount Gun."*

In continuation of iv. The gun will be dismounted and, together with the tripod and ammunition box, will be placed in the same position as at the beginning of Test i.

Points to be observed.

All the points as at the beginning of Test i. and the cork plug replaced. Standard time, 15 seconds.

vi. *To bring the gun into action on the command "Action" (object, range), e.g., "Action—fir tree—400."*

This test combines i. to iii. It emphasizes the necessity for proficiency in all details required before a gun can open fire with effect. This test should therefore not be applied until proficiency has been attained in each of those that precede it. The numbers, gun, tripod, and ammunition box will be as at the beginning of i.

Points to be observed.

All points as laid down for Tests i., ii., and iii. to be fulfilled. When No. 1 is ready to open fire, No. 2 will hold up his hand. Standard time, 40 seconds, from the range being ordered until No. 2 holds up his hand.

vii. *Horizontal traversing. On the command "(limits of traverse).........Traversing Fire."*

The target will be a horizontal line of figures, khaki on green, 3 inches high and 4 inches apart from centre to centre, placed at 25 yards from the gun. The gun will be laid on any figure that may be ordered, sights set at 500 yards. The test will comprise traversing from *right* to *left* as well as from *left* to *right*. On the command "Traversing Fire," No. 1 will fire a group at the figure named, then traverse, so that the next group will be fired at the interval to the next figure; the subsequent groups will be fired at a figure and a space alternately. Each time a group is fired, pressure will be maintained on the double button for approximately 1 second, which is about the time required to fire a group of 5 to 10 rounds. The test will not be completed until the space included between 5 figures, including the first-named and the fourth from it, has been traversed. In order to ensure that the traversing is satisfactory throughout, the order to cease fire will be given at least once during the traverse, but not before 5 groups have been fired, and the laying will be checked; this will be repeated when the limit of the traverse is reached.

Points to be observed.

That the traversing clamp is just sufficiently loose to enable the gun to be deflected by a sharp tap with the hand on the rear cross-piece; when checking the laying, that the sights are laid approximately correctly; the object is to test the traversing by ascertaining if the strength of tap has been correctly estimated, and not accurate re-laying; tapping backwards to obtain accuracy of aim will not be allowed. By counting the number of groups fired, the point of aim can be calculated, *e.g.*, fire opened on the first figure and stopped after the 7th group has been fired; the gun should then be laid on the fourth figure. Standard time, 3 seconds for each complete series, *i.e.*, a group and a completed traverse; *e.g.*, in the example above, the time taken should have been 19 seconds, *i.e.*, 6 complete series = 18 seconds; and a group = 1 second; total = 19 seconds.

viii. *Diagonal traversing. On the command* "*(Limits of traverse)........Traversing Fire.*"

The target will be three lines of 3 figures as for vii., each joined at an angle of 120 degrees to the next.

The procedure will be as for vii., but in this test correct manipulation of the elevating wheel is included. Traversing will be from *right* to *left* as well as from *left* to *right*.

Points to be observed.

As in Test vii. Standard time, 4 seconds for each complete series as explained in Test vii.

ix. *Rectifying stoppages.*

The instructor will indicate the stoppage required by adjusting the crank handle of a spare gun, if available, or by holding a stick against a wall or target to exemplify the position of the crank handle which he wishes to illustrate. For example:—Crank handle vertical; immediate action. The other positions of the crank handle can be similarly exemplified.

As an elementary test only the "immediate action," *see* Column 2 of Table of Stoppages, Machine-gun Handbook, will be required, but as proficiency increases the remedy of stoppages may be more fully tested by introducing variations in accordance with the tabulated list of stoppages, 5th column, "Remedy in Detail"; *e.g.*, after the "immediate action" in above example has been applied, keep the crank handle in the same position, telling No. 1 at the same time "Gun still stops." No. 1 should lighten the fuzee spring or put on the muzzle attachment. In all cases the "immediate action" must first be applied.

Points to be observed.

That the correct remedy is applied and completed; that all motions are correctly and clearly carried out; that the gun is re-laid correctly after a remedy has been completed. Standard time—The correct procedure to be begun within 3 seconds of the order "gun stops" from the instructor.

x. *Belt filling.*

(*a*) A heap of 25 rounds of ball ammunition to be placed beside a man; these to be inserted in a belt. Standard time, 1 minute.

(*b*) As for (*a*), but 250 rounds to be inserted in a belt by 1 man. Standard time, 12 minutes.

Points to be observed.

Rounds to be placed anyhow in a heap and not arranged. Inspection of the belt on completion will show if it has been filled so as not to cause a fault in feed.

Section 114.—*Machine gun course.*

643. Table "C" is divided into three parts:—

Part I is instructional, and since it is probably the first time a new machine gunner fires with service ammunition, careful and thorough instruction is necessary throughout the practices of Part I. The trained gunner must also regard these practices as instructional. The best value will be obtained by criticizing each practice while it is in progress, ceasing fire for the purpose, rather than by waiting until it is completed, when more ammunition will probably not be available with which to correct faults.

In these practices the firer learns, as he gains experience, to understand thoroughly the peculiarities of his gun and its mounting, and to compensate for them by suitable holding. These points can seldom be learnt without careful instruction and explanation by the section officer.

In the traversing practices of Part I no tapping backwards to correct faulty traversing will be allowed. In these practices a space exceeding 2 inches without a bullet mark indicates faulty traversing.

644. The sighting elevation to be used for the instructional machine gun target is 800 yards. Aim taken at or in line with the feet of a figure should cause the bullets to strike in the centre of the band vertically above. Guns should, however, be harmonized before firing. As the accuracy of the first shot of each group fired by a machine gun is not reliable, a wide shot will generally be found when examining a group. Instructors should bear this in mind when criticizing the results of a practice or measuring the size of a group. In single shot traversing, therefore, the elevation may differ slightly from that required when firing groups.

645. Practices of Part I may be repeated as considered necessary by the section officer, provided that the rounds allotted to Parts I and III are not exceeded.

646. In the classification practices (9, 10, 11, 12) of Part II, fire will be stopped as soon as the time limit is reached. No allowance will be made in these practices for stoppages which are due to causes other than defects of the mechanism or breakages. The firer will be given time to look over the gun and the ammunition belt before each practice is begun.

Should the stoppage be due to a defect in the mechanism or to a breakage, sufficient time to remedy such stoppage will be allowed, or the practice will be repeated.

647. Points will be allotted in the classification practices (9, 10, 11, 12) as follows:—

Practice 9.	Points.
75 per cent. of hits and over	35
60 " " and less than 75 per cent.	30
45 " " " " 60 "	25
30 " " " " 45 "	15
15 " " " " 30 "	5
Less than 15 "	0

Practice 11.	Points.
50 per cent. of hits and over	35
40 " " and less than 50 per cent.	30
30 " " " " 40 "	25
20 " " " " 30 "	15
10 " " " " 20 "	5
Less than 10 "	0

	Practice 10 (Points).	Practice 12 (Points).
No spaces	45	65
Not exceeding 2 spaces	40	60
" " 4 "	30	50
" " 6 "	15	35
" " 8 "	5	20
" " 10 " (exceeding 8 spaces = 0)		5
Exceeding 10 "		0

In Practices 10 and 12 the target should be divided into rectangles by means of invisible vertical lines 1 foot 6 inches apart. A space means any rectangle which does not contain a bullet mark. A hit on a dividing line to count as most favourable to the firer, but will only count in one rectangle.

In order to be classified as a "1st Class" Gunner, 100 points must be obtained in the classification practices.

In order to be classified as a "Qualified" Gunner, 50 points must be obtained in the classification practices.

Those who obtain less than 50 points will be classified as "Inefficient," and should generally be replaced in the Machine gun Section.

648. *Part* III. The ammunition allotted to this Part will be at the disposal of the section officer, and may be expended as he considers most beneficial for the efficiency of his machine gun section. No record of these rounds, other than the entry in the ammunition book, except such as are expended for extra practice of bad shots, need be kept, nor should any statement of expenditure be required.

Field practices.

649. *Part* IV. Tactical exercises, or problems in accordance with the principles laid down for field practices fired with the rifle, should be carried out. Most of these practices should be carried out between ranges of 600 and 1,200 yards.

650. *Allotment of ammunition.*

		Rounds.
Part I.	Each officer, non-commissioned officer and man	170
Part II.	Each non-commissioned officer and man ...	550
	Section officer	350
	Total	10,600

	Rounds.
Part III (*see* para. **648**)	265
To the Commanding Officer for expenditure as he may consider advisable, such as training of drivers and horse-holders (cavalry), field practices, tests of machine guns or personnel, experimental firing, demonstrations	1,900
Training of Reserve section, Part I	2,500
To the Brigadier, for each section of the Brigade, to be expended at his discretion	2,235
Total per section	17,500

Ammunition allotted to machine gun sections is not to be expended for rifle firing.

651. One officer, 2 non-commissioned officers, and 12 men per regiment and battalion will be trained as the Service machine gun section, and 2 non-commissioned officers and 12 men as the Reserve section, allotments of ammunition for each being as already detailed.

Machine Gunners of Royal Garrison Artillery (Special Reserve) allotted to armament guns of coast defences. Machine Gun Course.

652. Part I., Table "C" to be fired as laid down in paras. **643**, **644**, and pages 233 to 235, except that 10 rounds only will be allotted to practice 4 instead of 20 rounds. Total 100 rounds.

For classification, repeat Part I., Table "C" as modified, viz. :—

Practice.	Rounds	Conditions for Classification.
1	6	*1st class.*—3-in. group. One wide shot allowed. *2nd class.*—4-in. group One wide shot allowed.
2	7	*1st class.*—Not more than 2 spaces. 2nd *class.*—Not more than 3 spaces.
3	12 (6 to each group)	*1st class.*—3-in. group. One wide shot allowed; point of mean impact to be within the rectangle vertically above the figures indicated. *2nd class.*—As above, except a 4-in. group and one wide shot allowed.
4	...	Not included in classification.
5	50	As for Practice 2.
6	15 (Single shots)	*1st class.*—Not more than 6 spaces. *2nd class.*—Not more than 8 spaces.

Ammunition—	Rounds.
Instructional	100
Classification	90
Surplus, at machine gun officer's disposal	10
Total	200

Notes.—1. A space means a horizontal interval not greater than 3 inches and not less than 1 inch between bullets in Practices 2 and 6 or between the nearest bullets of adjacent groups in Practice 5.

2. In Practice 2 (classification practices) it is not necessary that the shots should be in the band.

3. In Practice 6 (classification practices) shots to count must be within the diagonal bands.

4. In Practices 2 and 6, tapping backwards to correct faulty traversing will not be allowed.

5. When repeating Class I. for classification, a practice will be completed by the firer without any assistance or criticism until its conclusion.

Table " C."—Special Reserve and Territorial Force.

653. Officers, non-commissioned officers, and men of the machine gun sections of the Special Reserve (except Royal Garrison Artillery) and Territorial Force will fire Part I, Table " C." Surplus ammunition may be used for repetition, or for such practices of Part II as the commanding officer may consider desirable.

653A. Regulars, Special and Extra Reserve, except Royal Garrison Artillery and Territorial Force, non-commissioned officers and men of detachments allotted to coast defence machine guns will carry out the following courses :—

(i.) *1st and 2nd years as machine gunners*—Part I., Table " C."

(ii.) *3rd and subsequent years*—

25 rounds per man will be fired in practices 1, 2, and 3, Part I., Table " C."

150 rounds will be fired in practical exercises to be framed by coast defence commanders against such targets as may have to be engaged in war.

The surplus 25 rounds will be at the disposal of

coast defence commanders for repetition of indifferent shots, ranging bursts, &c.

Instructions.

(*a.*) Each man of Regular Artillery and Infantry of the machine-gun detachment detailed to armament machine guns, in addition to firing the annual course, will be trained for at least 2 days in every month with the guns that would be used on mobilization.

There is no objection in cases where it may be more convenient and advisable to detail the machine-gun detachments for 24 consecutive days' practice instead of 2 days monthly.

(*b.*) *Special and Extra Reserve and Territorial Force.*—The training of the machine gunners is not to aim at a high tactical standard. The short time available for training should therefore be devoted to preliminary work and in giving thorough instruction in the mechanism and manner of firing the weapon.

The training should, as far as possible, be as follows :—

Handbook, ·303-inch Maxim Gun, Part I.

Infantry Training, Section **9**, paragraphs 1 and 3 to 7 inclusive. Sections **114** to **117** inclusive, and Section **166**.

Musketry Regulations, Part I. (reprint), paragraphs **279** and **280** ; Chapter V., paragraphs **641** to **645**, first line only.

(*c.*) *Special and Extra Reserve.*—The firing of the annual course, if possible, will not be carried out until after the execution of the annual rifle (musketry) course.

(*d.*) *Territorial Force.*—Machine gunners detailed to armament coast defence guns will, in addition to firing their annual machine-gun course, be exercised as frequently as possible with the guns which they would be required to man on mobilization.

NOTE.—Machine gunners allotted to coast defences (except men of the Royal Garrison Artillery of the Territorial Force), will also fire their annual course of musketry. They will not be classified on their firing with armament machine guns.

Table "C."

MACHINE GUN COURSE (ANNUAL).

Cavalry and Infantry—Regular Army.

PART I.—INSTRUCTIONAL.

To be fired at a range of 25 yards. Target, Instructional Machine gun Target, Plate 35, Musketry Instruction, Part II.

No.	Nature of Practice.	Rounds.	Method of Conducting and Object of Practice.
1	Grouping	6	To teach the importance of the correct holding required for the gun, which should group in a 8-inch ring.
2	Single shot Traversing	7	To teach automatic tapping. Four figures to be indicated by the instructor. Gun to be laid on the flank figure indicated by the instructor; fire a shot and tap alternately as in Section Drill (Traversing Fire). Shots should be approximately 2 inches apart. The result of each shot should be criticized. Single shot loading.
3	Application	12 (6 to each group)	To teach correct laying and holding. Two alternate figures to be indicated by the instructor. A group to be applied to the rectangle above each figure. The point of mean impact of each group should be within the rectangle above each figure respectively.

Table "C"—continued.

No.	Nature of Practice.	Rounds.	Method of Conducting and Object of Practice.
4	Vertical Searching	20	To teach automatic manipulation of the elevating wheel. Single shot loading. The gun to be laid on a figure with sights adjusted to 800 yards. Without altering the elevation of the gun, adjust the sights to 1,250 yards. Fire a shot, then elevate and fire, and continue elevating and firing alternately until the sights are again aligned on the original aiming mark. Each shot should be approximately 2 inches vertically above the last, and the thin brown band should be reached. Then traverse about 2 inches inwards and, without altering the elevation of the gun, adjust the sights to 800 yards and proceed as before, but depressing after each shot instead of elevating. When the sights are aligned between the figure originally laid upon and the next, the practice is completed, and each shot should be approximately 2 inches vertically below the last. The vertical interval of 2 inches at 25 yards is the horizontal equivalent to 60 yards at 1,000 yards range, or about the depth of the effective zone for the range.
5	Traversing	50	Having learned to know the holding required for the gun in Practices 1 and 3, and Practice 2 having afforded practice in automatic tapping, instruction is now given in practical traversing

Table "C"—continued.

No.	Nature of Practice.	Rounds.	Method of Conducting and Object of Practice.
			by groups of 5 to 6 rounds. Five figures to be indicated by the instructor. Gun to be traversed from *right to left*. Groups should be evenly distributed along the band above and between the figures indicated; there should be no space exceeding 2 inches without a bullet mark.
6	Diagonal Traversing	75	To teach manipulation of the elevating wheel combined with traversing. Gun to be traversed from the second figure from the left to the second figure from the right of the three bands. The same principles hold good as in Practice 4.
	Total rds. per man	170	

Table "C"—continued.

PART II.—CLASSIFICATION PRACTICES 9, 10, 11, 12 ONLY.

No.	Nature of Practice.	Target Screen (covered with Regulation brown paper).	Range (Yds.).	Rds.	Time (Secs.).	Remarks.
7	Ranging	3′ high 10′ wide	400	50	—	In these practices each man has an opportunity of sighting his gun on the open range before firing the classification practices.
8	Ranging	"	600	50	—	In these practices each man has an opportunity of sighting his gun on the open range before firing the classification practices.
9	Application	"	400	50	20	
10	Traversing	3′ high 30′ wide	400	100	50	Gun to be traversed from *right to left*. The firer is required to traverse the target with the rounds allotted within the time limit without restrictions.
11	Application	3′ high 10′ wide	600	50	20	
12	Traversing	3′ high 30′ wide	600	100	50	Gun to be traversed from *left to right* under the same conditions as in Practice 10.

Table "C"—continued.

No.	Nature of Practice.	Target Screen (covered with Regulation brown paper).	Range (Yds.).	Rds.	Time (Secs.).	Remarks.
13	Observation	3′ high 10′ wide	900 to* 1200	100	†	Range known approximately. If it is probable that the firer can himself observe, he should apply his fire from such observation. The remainder of the section, except a No. 2 to assist the firer, should form two groups under the serjeant and corporal respectively. These groups should observe the fire by eye or with field glasses from the flanks. Each N.C.O. and man should note down the result of his observation of each group fired, and at the end of the practice put against each note of his observation the semaphore signal he would have sent had he been required to signal results. If the firer is unable to obtain observation himself, another No., not the

* According to range facilities, nature of ground and climatic conditions.

† No limit, but at a rate of at least 250 rounds a minute.

Table "C"—continued.

No.	Nature of Practice.	Target Screen (covered with Regulation brown paper).	Range (Yds.).	Rds.	Time (Secs.).	Remarks.
						No. 2 at the time, will control the fire from observation with field glasses, the remainder observing from a flank as described above. After each firer has completed the practice, the section officer will criticize the results of the observation as regards methods followed by firer and observers. During this practice the belts may be prepared with artificial stoppages placed after every 20 or 30 rounds in the belt. The time and method required to remedy them should be noted and criticized.
14	Fire from successive positions.	15 iron falling plates on a frontage of 30 feet.	800 to 400	100	—	Ranges known approximately. Nos. 1, 2 and 3 will fall in with the gun, tripod and ammunition box, as for the 1st Test of Elementary Training, about 100 yards in rear of first fire position, which will be approximately 800 yards from the targets.

Table "C"—continued.

No.	Nature of Practice.	Target Screen (covered with Regulation brown paper).	Range (Yds.).	Rds.	Time (Secs.).	Remarks.
						The section officer will mark the first and subsequent fire positions, and, on his signal, the gun, &c., will be carried forward at a steady double, and fire opened without further orders. The gun will be carried dismounted, and the tripod legs will be closed and clamped until the fire position is reached. Fire will be continued at each position until a hit is obtained. Sights will not be adjusted until the new position is reached. At alternate fire positions the firer will adopt the prone position when firing. The objects of this practice are to emphasize the lessons of elementary training as regards quick and correct mounting of the gun and quick opening of fire, and also to exemplify the principle of maintaining fire until effect is obtained. It is often desirable to time the practice or a

Table "C"—continued.

No.	Nature of Practice.	Target Screen (covered with Regulation brown paper).	Range (Yds.)	Rds.	Time (Secs.)	Remarks.
						portion of it. If the section officer decides to do so, the time should be taken until a hit is obtained—not merely until fire is opened, because this encourages men to open fire without accurate laying. The advance to a fire position should not be a race between detachments if both are carrying out the practice simultaneously, nor against time.

CHAPTER XI.

SCHOOL OF MUSKETRY.

Section 116.—*Object of the school.*

651. The School of Musketry is established for the following purposes:—

i. To train officers, warrant and non-commissioned officers to act as instructors in musketry and the machine gun. To teach the principles of fire tactics. To afford practical demonstration of the power of fire, and by means of experimental firing and object lessons to illustrate methods of employment of fire and their value under varying conditions. To enable officers to obtain a sufficient knowledge of the scientific side of musketry, and of foreign arms and musketry regulations.

ii. To carry out such trials of small arms and ammunition as the Army Council may require, and such other trials and experiments as may be necessary for the instruction of the Army, and to advance the knowledge of small arms and of musketry generally.

iii. To study all questions connected with small arms, ammunition, and fire tactics; to examine and develop musketry inventions; and to adapt musketry appliances to the needs of the Army.

iv. To study and keep touch with the regulations, arms, methods, and progress in foreign armies.

v. To study the question of range apparatus, and to submit drawings and plans of ranges and range appliances.

Section 117.—*The commandant.*

655. The commandant will submit to the Army Council extracts from the reports from general officers commanding, together with a compilation of the results of the annual course of the troops serving at home, his own report, and recommendations on all matters which concern musketry training.

656. He will direct the instruction at the School of Musketry, and will furnish a yearly report thereon to the Army Council.

657. He will transmit yearly to the Army Council results of experimental firing carried out at the School.

Section 118.—*Courses of instruction.*

658. Courses of instruction in musketry will be held as follows :—

i. *Qualifying rifle courses.*

(*a*) For officers, warrant and non-commissioned officers of the Regular Forces.

(*b*) For officers and non-commissioned officers (other than the Regular Establishment and permanent staff) of the Special Reserve and Territorial Force.

ii. *Qualifying machine gun courses.*—For officers and non-commissioned officers of all branches of the service.

iii. *Qualifying range-taking courses.*—For selected officers of the Regular Forces.

iv. *Refresher courses.*—Rifle and machine gun.

v. *Senior officers' courses.*

vi. *Special courses as may be ordered.*

659. *Cancelled.*

660. Attendance at the School of Musketry is governed by the King's Regulations.

661. For all courses the dates mentioned in Army Orders are the dates appointed for the *assembly* and *dispersal* of classes. Officers, warrant and non-commissioned officers ordered to attend should report their arrival at the School before 7 p.m. on the day of assembly.

662. The detachment pay sheets of warrant and non-commissioned officers will be sent by post to the adjutant so as to arrive by the date on which the class assembles.

663. Non-commissioned officers of the Special Reserve and Territorial Force will be dealt with as regards pay and allowances in accordance with the regulations for their respective branches.

664. Officers should be provided with warrants for the outward and homeward journeys. Warrant and non-commissioned officers will be furnished with return routes and separate warrants for the homeward journey, unless it is known that they will not return to the same station. In the case of serjeants of the regular establishment of the Special Reserve and permanent staff of the Territorial Force who are required to proceed to the Royal Small Arms Factory, Enfield Lock, on completion of the course, the route should provide for the return journey, *viâ* Enfield Lock, warrants being issued accordingly.

Section 119.—*Record of qualifications.*

665. Certificates are not issued. The commandant will forward to the War Office lists of the officers who have attended, showing the results of their examination, and will furnish general officers commanding and the Deputy Adjutant General Royal Marines with the names of officers, warrant and non-commissioned officers who have qualified, and, confidentially, with the names of those who have failed. The names of those who have qualified will be published in Command or District Orders, and the qualification entered in records of service. It should be specially noted on the record of service whether the qualification is in the rifle, the Maxim gun, or range-taking. Those officers, warrant and non-commissioned officers who attain a sufficient standard will be noted as "*Distinguished.*"

666. The names of officers and non-commissioned officers of the Royal Horse and Royal Field Artillery, and of heavy batteries, Royal Garrison Artillery, who qualify as brigade instructors in musketry, of officers of companies Royal Garrison Artillery, of officers and non-commissioned officers of the Army Service Corps, and of officers and non-commissioned officers of the Army Ordnance Corps, who qualify as instructors in musketry with infantry battalions, will be published in command or district orders, and the qualification entered in the record of service. General officers commanding will submit results, in duplicate, as early as possible after the termination of each examination, to the commandant, School of Musketry, Hythe, who will transmit one copy to the War Office.

667. Officers, warrant and non-commissioned officers whose companies execute the range practices of the annual course during their attendance at Hythe, or any other School of Musketry, may, at the discretion of officers commanding, be recorded in Company Annual Returns as "Exercised at a School of Musketry," but will take part in all musketry exercises carried out after they have rejoined. Those who, owing to attendance at a course, are not classified by their units, may, if in possession of marksmen's badges, continue to wear them until again classified. Transfer returns will not be rendered from Hythe.

Section 120.—*Serjeant-instructors.*

668. Candidates for these appointments must be serjeants or lance-serjeants who have qualified at a school of musketry. Application will be made on Army Form B 241 to the Staff Officer, School of Musketry, Hythe, and copies of the company conduct sheet and record of service of each candidate must accompany the application.

APPENDIX I.

CAVALRY, R.E., AND INFANTRY—TABLES A AND B.

Table "A."—Recruits' Course—Regular Forces (Cavalry, Royal Engineers, and Infantry), and Special Reserve (Royal Engineers and Infantry).

N.B.—Royal Engineers (Sappers and Pioneers) of the Regular Forces and Royal Engineers Special Reserve will fire those practices only which are detailed in the Instructions for Royal Engineers. (*See* Section 87.)

Recruit drivers of the Royal Engineers will fire:—Part I.—Practices 1, 2, 3, and 4. Part II.—Practices 6, 8, and 11. Part III.—Practice 20.

The standards prescribed in Practices 6, 8, and 11 should be reached before proceeding to Practice 20. Surplus ammunition may be expended in repetition. Those who obtained 65 points in Practices 3, 4, 6, 8, 11 and 20, will be classified as 2nd Class.

No.	Practice.	Target.	Distance in yards.	Rounds.	Instructions for Conduct of Practice.
		PART I.—INSTRUCTIONAL PRACTICES (ELEMENTARY).			
1	Grouping ...	2nd Class Elementary (Bull's-eye)	100	5	Lying, with arm or rifle rested.
2	Application ...	2nd Class Elementary (Bull's-eye)	200	5	Lying, with arm or rifle rested.
3	Grouping ...	2nd Class Elementary (Bull's-eye)	100	5	Lying.
4	Application ...	2nd Class Elementary (Bull's-eye)	200	5	Lying.
		Total rounds ...		20	

No.	Practice.	Target.	Distance in yards.	Rounds.	Positions and Grouping Standards.
		PART II.—INSTRUCTIONAL PRACTICES (REPETITION).			
5	Grouping ...	2nd Class Elementary (Bull's-eye)	100	5	Lying. All shots in 12-inch ring.
6	Application ...	2nd Class Figure ...	200	5	Lying, with arm or rifle rested. Five hits, including 4 within Inner (24-inch) ring.
7	" ...	" " ...	200	5	Lying. Five hits within Magpie (36-inch) ring.
8	" ...	" " ...	300	5	Lying. Five hits.
9	" ...	1st Class Figure ...	200	5	Kneeling. Four hits at least within Inner (40-inch) ring.
10	" ...	" " ...	300	5	Kneeling with arm or rifle rested. Four hits at least within Inner (40-inch) ring.
11	" ...	" " ...	400	5	Lying. Four hits at least.
12	" ...	" " ...	500	5	Lying, with arm or rifle rested.
13	" ...	" " ...	500	5	Lying.
14	" ...	" " ...	600	5	Lying, with side of rifle only rested
		Total rounds ...		50	

No.	Practice.	Target.	Distance in yards.	Rounds.	Instructions for Conduct of Practice.
		PART III.—INSTRUCTIONAL PRACTICES (TIMED).			
15	Slow	2nd Class Figure ...	200	5	Lying.
16	,, ...	,, ,, ...	200	5	Kneeling.
17	Rapid ...	,, ,, ...	200	5	Lying. 40 seconds allowed.
18	Slow	1st Class Figure ...	400	5	Lying.
19	Rapid ...	,, ,, ...	400	5	Lying. 40 seconds allowed.
20	Slow	,, ,, ...	500	5	Lying. Taking cover behind stones or sandbags representing a parapet and firing over them.
21	Snapshooting	2nd Class Figure ...	200	5	Lying. Exposure, 6 seconds for each shot.
22	,,	,, ,, ...	200	5	Kneeling. Taking cover in a trench, or behind a screen representing a wall, and firing over the parapet. Exposure, 6 seconds for each shot.
		Total rounds ...		40	

Part IV.—Instructional Practices (Classification for Special Reserve).

23	Grouping ...	2nd Class Elementary (Bull's-eye)	100	5	Lying.
24	Application ...	1st Class Figure ...	300	5	Kneeling.
25	Rapid ...	" " ...	300	5	Lying. 40 seconds allowed.
26	Snapshooting	2nd Class Figure ...	200	5	Lying. Taking cover as in 20. Exposure, 5 seconds for each shot.
27	Application ...	1st Class Figure ...	500	5	Lying.
		Total rounds ...		25	

Part V.—Individual Field Practices.

Twenty rounds will be expended in elementary practices, 10 rounds in an attack practice from 700 to 200 yards, and 10 rounds in a defence practice against full-length figures representing an advancing enemy.

Total rounds	20

Part VI.—Collective Field Practices.

Twenty-five rounds will be expended, if ammunition is available.

Total rounds	25
Surplus rounds	20
Total rounds for Table A ...	200

Table "B."—Annual Course—Regular Forces (Cavalry, Royal Engineers, and Infantry).

N.B.—The Royal Engineers (Sappers, Regular Forces), including Regular Establishment of the R.E. Special Reserve, will fire those practices only which are detailed in the Instructions for Royal Engineers. (*See* Section 87.)

Trained drivers of the Royal Engineers will fire :—

Part I.—Practices 1, 2, 3, 5 and 6.
Part II.—Practices 7 and 9.
Part III.—Practice 19.

They will be classified on the same standard as laid down for the 40 rounds fired in Table B, Appendix II, page 260.

No.	Practice.	Target.	Distance in yards.	Rounds.	Instructions for Conduct of Practice.
		PART I.—QUALIFYING PRACTICES.			
1	Grouping ...	2nd Class Elementary (Bull's-eye)	100	5	Lying, with arm or rifle rested.
2	Application	" " "	200	5	As in 1.
3	"	2nd Class Figure ...	300	5	Kneeling, with arm or rifle rested.
4	Grouping ...	2nd Class Elementary (Bull's-eye)	100	5	Lying.
5	Application	1st Class Figure ...	400	5	Lying.
6	"	" " ...	500	5	Lying, with side of rifle only rested.
		Total rounds ...		30	

PART II.—INSTRUCTIONAL PRACTICES (TIMED).

7	Snapshooting	2nd Class Figure ...	200	5	Lying. Taking cover behind stones or sandbags representing a parapet and firing over them. Exposure, 6 seconds for each shot.
8	,,	,, ,, ...	200	5	Sitting or kneeling. Bayonet fixed. Exposure, 6 seconds for each shot.
9	Rapid	,, ,, ...	200	5	Lying. Bayonets fixed. 30 seconds allowed.
10	Slow	,, ,, ...	300	5	Lying.
11	Rapid	,, ,, ...	300	10	Lying. Rifle unloaded and magazine empty until the target appears. Loading from the pouch or bandolier by 5 rounds afterwards. One minute allowed.
12	Slow	1st Class Figure ...	500	5	Lying.
13	Rapid	,, ,, ...	500	5	Lying. Taking cover as in 7. 45 seconds allowed.
14	Slow	,, ,, ...	600	5	Lying. Taking cover behind stones or sandbags and firing round them, with side of rifle only rested.
15	Snapshooting	Figure No. 3 (silhouette)	200	5	Lying. Taking cover as in 14. Exposure, 4 seconds for each shot.
16	,,	,, ,,	200	5	Kneeling. Taking cover in a trench or behind a screen representing a wall and firing over the parapet. Exposure, 5 seconds for each shot.

No.	Practice.	Target.	Distance in yards.	Rounds.	Instructions for Conduct of Practice.
		Part II.—Instructional Practices (Timed)—*continued.*			
17	Crossing shot	Figure No. 6 (silhouette)	200	5	Lying. One shot at each run of 30 feet. Pace of target—quick time.
		Total rounds ...		60	
		Part III.—Classification Practices.			
18	Grouping ...	2nd Class Elementary (Bull's-eye)	100	5	Lying.
19	Snapshooting	Figure No 3 (silhouette)	200	5	Lying. Taking cover as in 7. Bayonet fixed. Exposure, 4 seconds for each shot.
20	Slow	2nd Class Figure ...	400	5	Lying. Taking cover as in 14.
21	,, ...	,, ,, ...	300	5	Kneeling. Taking cover as in 16.
22	Rapid	,, ,, ...	300	15	Lying. Rifle to be loaded and 4 rounds in the magazine before the target appears. Loading from the pouch or bandolier by 5 rounds afterwards. One minute allowed.
23	Slow	1st Class Figure ...	500	5	Lying.
24	Rapid	,, ,, ...	500	5	Lying. 30 seconds allowed.
25	Slow	,, ,, ...	600	5	Lying. Taking cover as in 7.
		Total rounds ...		50	

Part IV.—Individual Field Practices.

Total rounds ...	35

Part V.—Fire Direction Practices.

Short series of shots will be fired at distances beyond 600 yards by officers and non-commissioned officers for practice in observation of fire, estimating atmospheric influences, and verifying sighting by trial shots. Screens, or any visible objects such as might serve as range marks on service, will be used as targets. About 300 rounds, drawn from the surplus, should suffice. Special fire direction exercises should be substituted for these practices if range accommodation does not extend beyond 600 yards.

Part VI.—Collective Field Practices.

Total rounds ...	50
Surplus rounds ...	25
Total rounds for Table B	250

APPENDIX II.

ROYAL ARTILLERY, ARMY SERVICE CORPS, AND ARMY ORDNANCE CORPS.

i. Preliminary instruction in the Royal Artillery, in the Army Service Corps, and Army Ordnance Corps may, owing to the limited time available for musketry instruction, be limited to aiming and firing instruction, judging distance and making allowance for wind at distances up to 800 yards, discernment and recognition of targets, choice of targets, regulation of the rate of fire, mutual assistance in ranging and observation of results.

ii. Much of the preliminary training will necessarily be given in broken periods, though at least five days' consecutive instruction for four hours daily should immediately precede the firing of the courses for recruits and trained soldiers.

iii. It will be necessary to discriminate between men in various stages of proficiency, so that more time may be devoted to those who need it, and less time to the more advanced.

iv. This discrimination will be based on the results of the tests of elementary training held from time to time; and it should be the object of instructors to bring every man up to standard before the five days' preliminary training begins.

v. Marksmen of the previous year may be excused from the preliminary training at the discretion of commanding officers.

vi. The standards prescribed in Sections 63 and 64 must be attained at a preliminary test before proceeding to the elementary course (Table A).

vii. Those who fail to qualify in the preliminary test, whether in their first or subsequent years, will either be put back for further instruction and be again tested on a later occasion, or will fire ten rounds only in grouping practices and be classified as 3rd Class shots.

viii. Those who pass the preliminary test will, if they are recruits or 3rd Class shots of the previous year, fire the elementary course (Table A); if they are 2nd Class shots, or better, they will fire the trained soldiers' course (Table B). A trained soldier who fails to reach the standard for 2nd Class, in Table B, will be relegated to the 3rd Class, and will fire Table A, provided sub-para. vi has been complied with, in the next or subsequent years until he again reaches the 2nd Class.

ix. Men who have passed out of the 3rd Class in Table A will not fire Table B during the same musketry year.

x. Non-commissioned officers and men who are classified as 1st Class shots in two consecutive years may be permitted in the second year, if ammunition is available, to fire, in addition to Table B, the classification practices of Part III, Table B, for cavalry and infantry, with a view to qualifying as marksmen. In this case application will be made to general officers commanding for special arrangements to be made, as in the case of cavalry and infantry, for independent supervision and marking.

Those who fail in these classification practices to attain the standard for marksmen will be classified as 1st Class shots.

Table "A."—Recruits' Course—Royal Artillery, Army Service Corps (Regulars and Special Reserve), and Army Ordnance Corps.

GROUPING PRACTICE FOR THOSE RECRUITS OR TRAINED MEN WHO FAIL IN THE PRELIMINARY TEST.

No.	Practice.	Target.	Distance in yards.	Rounds.	Position.
1	Grouping ...	2nd Class Elementary (Bull's-eye)	100	5	Lying, with rest.
2	Grouping ...	2nd Class Elementary (Bull's-eye)	100	5	Lying.
		Total rounds...		10	

Elementary Course for 1st Year Men and 3rd Class Shots.

No.	Practice.	Target.	Distance in yards.	Rounds.	Position and Grouping Standard.
1	Grouping ...	2nd Class Elementary (Bull's-eye)	100	5	Lying. All shots in 12-inch ring.
2	Application...	2nd Class Elementary (Bull's-eye)	200	5	Lying, with rest. 5 hits, including 4 within Inner (24-inch) ring.
3	Application...	2nd Class Elementary (Bull's-eye)	200	5	Lying. 5 hits, including 4 within Inner (24-inch) ring.
4	Application...	1st Class Figure ...	300	5	Lying. Taking cover behind stones or sandbags representing a parapet and firing over them. 4 hits.
5	Application...	1st Class Figure ...	400	5	Lying.
6	Application...	1st Class Figure ...	500	5	Lying. Taking cover as in 4.
		Total rounds...		30	

The course should be fired in five days as follows:—

Practices 1 and 2, *first day*. Repetition, *second day*. Practices 3 and 4, *third day*. Repetition, *fourth day*. Practices 5 and 6, *fifth day*.

Practices 1, 2, 3 and 4 will be repeated once each if the grouping standard is not attained, but not on the same day. Should the firer have attained the grouping standard in any of the first four practices without repetition, he may on conclusion of all the practices be given the option of repeating No. 6 with a view to increasing his score for qualification.

In all cases of repetition the second score only will count for classification.

As a rule five days should be devoted to firing the course, of which the second and fourth would be reserved for repetition and careful instruction of indifferent shots.

Surplus rounds will be used as the officer commanding the battery or company may direct on preliminary or further training, testing rifles, occasional shots, or marksmen's firing in Part III, Table B, for cavalry and infantry.

Points required for 2nd Class = 65. Grouping standards have no effect on classification; they govern repetition only.

Table "B."—Annual Course—Royal Artillery, Army Service Corps and Army Ordnance Corps.

TRAINED SOLDIERS' COURSE FOR MARKSMEN, 1ST AND 2ND CLASS SHOTS.

No.	Practice.	Target.	Distance in yards.	Rounds.	Position.
1	Grouping ...	2nd Class Elementary (Bull's-eye)	100	5	Lying.
2	Application ...	2nd Class Figure ...	200	5	Lying. Taking cover behind stones or sandbags and firing round them with side of rifle only rested.
3	Application ...	2nd Class Figure ...	300	5	Sitting or kneeling.
4	Application ...	1st Class Figure ...	400	5	Lying.
5	Rapid... ...	1st Class Figure ...	400	5	Lying. 40 seconds allowed.
6	Application ...	1st Class Figure ...	500	5	Lying. Taking cover behind stones or sandbags representing a parapet and firing over them.
7	Snapshooting	2nd Class Figure ...	200	5	Lying. Taking cover as in 2. 5 seconds exposure.
8	Observation...	Iron Falling	300	5	Lying. Taking cover as in 6. Firing in pairs.
		Total rounds...		40	

Three points will be allowed for every direct hit in Practice 8.

In Practices 5 and 8 there will be no signalling until the firer has completed his practice.

In Practice 8 there will be separate targets for each firer, and the firers in each pair will fire alternately, assisting each other by observation.

Classification :—

Those who obtain 95 points or more ... 1st Class shots.
Those who obtain 60 points and less than 95... ... ... 2nd Class shots.
Those who obtain less than 60 points ... 3rd Class shots.

APPENDIX III.

SPECIAL RESERVE, TABLES A AND B.

Recruits.

The practices to be fired by recruits of the Special Reserve infantry are the same as those prescribed for recruits of the cavalry and infantry of the Regular Army, but the ammunition allowed for Part I will be limited to 20 rounds.

Recruits who fail to obtain 90 points in Part II will be classified as 3rd Class shots and repeat the whole of Part II before proceeding to Part III; they will not fire Parts IV and VI. Those who obtain 90 points in Part II, but fail to reach the Grouping Standard in any practices of Part II, will repeat once each those practices (not exceeding 5 in number) in which they failed before proceeding to Part III, the rounds to be expended in Part VI being in their case reduced by a corresponding amount.

Recruits who fire Part IV will be classified upon their total score in Part IV as follows:—

Those who obtain 80 points and upwards ... 1st Class shots.
" " 40 " less than 80 points ... ... 2nd Class shots.
" fail to obtain 40 points ... ... 3rd Class shots.

Recruits of the Royal Engineers (Special Reserve) will fire Parts I, II, III and IV of the same table, but will not repeat Part I nor any practices of Part II unless they fail to make a total of 90 points in Part II.

Those who fail to obtain 90 points will be classified as 3rd Class shots, and repeat Practices 5 to 12, instead of firing Part IV.

Those who fire Part IV will be classified according to the standards prescribed for infantry (Special Reserve).

Surplus ammunition will be used for additional practice of indifferent shots.

Table "B."—Annual Course—Special Reserve (Supplementary Officers and Infantry).

No.	Practice.	Target.	Distance in yards.	Rounds.	Instructions for Conduct of Practice.
		PART I.—QUALIFYING PRACTICES.			
1	Grouping ...	2nd Class Elementary (Bull's-eye)	100	5	Lying.
2	Application ...	2nd Class Figure ...	200	5	Lying.
3	Snapshooting	" " ...	200	5	Lying. Taking cover behind stones or sandbags representing a parapet and firing over them. Exposure, 6 seconds for each shot.
4	Application ...	1st Class Figure ...	300	5	Kneeling. Taking cover in a trench or behind a screen representing a wall and firing over the parapet.
5	Rapid ...	" " ...	300	5	Lying. 35 seconds allowed.
6	Application ...	" " ...	500	5	Lying. Taking cover behind stones or sandbags and firing round them, with side of rifle only rested.
		Total rounds...		30	

No.	Practice.	Target.	Distance in yards.	Rounds.	Instructions for Conduct of Practice.
		PART II.—CLASSIFICATION PRACTICES.			
7	Grouping ...	2nd Class Elementary (Bull's-eye)	100	5	Lying.
8	Snapshooting	Figure No. 3 (silhouette)	200	5	Lying. Taking cover as in 3. Bayonet fixed. Exposure, 4 seconds for each shot.
9	Slow	2nd Class Figure ...	400	5	Lying. Taking cover as in 6.
10	"	" " ...	300	5	Kneeling. Taking cover as in 4.
11	Rapid ...	" " ...	300	15	Lying. Rifle to be loaded and 4 rounds in the magazine before the target appears. Loading from the pouch or bandolier by 5 rounds afterwards. 1 minute allowed. (The scoring will be as for Practice 22, Table B, Appendix I. *See* Note 2 to para. **495**.)
12	Slow	1st Class Figure ...	500	5	Lying.
13	Rapid ...	" " ...	500	5	Lying. 30 seconds allowed.
14	Slow	" " ...	600	5	Lying. Taking cover as in 3.
		Total rounds...		50	

Trained soldiers (Special Reserve, infantry) will not fire Table B in the same year as that in which they fire Table A.

Those who fail to obtain 50 points in Part I., Table "B," should receive further instruction with dummy cartridges and miniature ammunition, and, if service ammunition is available, will fire 20 additional rounds in selected instructional practices before proceeding to Part II.

Men will be classified on their performance in Part II. only, and in accordance with the following standards :—

Those who obtain 130 points and upwards	Marksmen.
Those who obtain 95 points and less than 130	1st class shots.
Those who obtain 65 points and less than 95	2nd class shots.
Those who obtain under 65 points	3rd class shots.

Recruits of the Army Service Corps (Special Reserve), will fire the course laid down for recruits of the Royal Artillery (Regulars).

The Irish Horse fire the same courses as the Yeomanry (Territorial Force).

Supplementary officers of the Special Reserve (cavalry and infantry) will fire Table B, Appendix III.

APPENDIX IV.

TERRITORIAL FORCE.

*Table "A."—Recruits' Course—Irish Horse, Territorial Force (Yeomanry, *R.E. Field Companies, and Infantry), the King's Own Malta Regiment of Militia, and R.E. (Militia), Malta Division.*

PART I.—INSTRUCTIONAL PRACTICES.

No.	Practice.	Target.	Distance in yards.	Rounds.	Instructions for Conduct of Practice.
1	Grouping ...	2nd Class Elementary (Bull's-eye)	100	5	Lying, with rest.
2	Application ...	"	200	5	" "
3	" ...	"	200	5	Lying
4	" ...	1st Class Elementary (Bull's-eye)	300	5	Lying, with rest.
5	" ...	"	300	5	Kneeling.
6	" ...	"	500	5	Lying, with rest.
		Total	...	30	

* Recruit Drivers will carry out Part I and Practices 9 and 10 of Part II only. The standard for qualification will be eliminated in the case of these drivers.

TERRITORIAL FORCE.

Table "A."—Recruits' Course—Territorial Force.

PART II.—STANDARD TEST.

No.	Practice.	Target.	Distance in yards.	Rounds.	Instructions for Conduct of Practice and Standards.
7	Grouping ...	2nd Class Elementary (Bull's-eye)	100	5	Lying. Qualification standard. Four shots in 12-inch ring, 5th shot not necessarily on the target. To count points, instructions in para. **443**, iii. and v., will be followed.
8	Application ...	"	200	5	Lying. Four hits, minimum score of 12 points.
9	Snapshooting	1st Class Elementary (Bull's-eye)	200	5	Lying. Taking cover behind stones or sandbags and firing round them, with side of rifle only rested. Exposure, 6 seconds for each shot. Four hits on target.
10	Application ...	"	500	5	Lying, with rest.
		Total rounds...	...	20	

Recruits will fire Table A in their first year. Those who fail to qualify* will fire this table in subsequent years until the qualification standard is reached.

The practices in Part II are not to be substituted for practices in Part I, and record must be kept of the practices fired by every officer, non-commissioned officer and man, with the results of their firing.

Except in repetition practices, not more than 20 rounds or the equivalent of 20 rounds should be fired in one day.

Instruction will be allowed during the firing of Part II.

The ammunition will be allotted as follows:—

Part I	30 rounds.
Part II	20 rounds.
Repetition, field practices, voluntary practices, and competitions at discretion of commanding officer	40 rounds.
Total	90 rounds.

Those who satisfy the standards in Part II and score 45 points will be classified as having reached the "qualification standard."

Any two practices of Part II may be repeated once each in order to reach the standard. In cases of repetition, the second score only will count for classification.

* The word qualify in these regulations means that the necessary standard for advancement to a further course of instruction has been attained. It is not to be taken as meaning qualification for pay or for earning grants or allowances.

*Table "B."—Annual Course—Irish Horse, Territorial Force (Yeomanry, *R.E. Field Companies and Infantry), and the King's Own Malta Regiment of Militia.*

N.B.—Trained men of the R.E. (Militia) Malta Division, will fire annually Practices 1, 3, 5 and 6 of Table "A," in lieu of Table "B."

PART I.—INSTRUCTIONAL PRACTICES.

No.	Practice.	Target.	Distance in Yards.	Rounds.	Instructions for Conduct of Practice.
1	Grouping ...	2nd Class Elementary (bull's-eye)	100	5	Lying.
2	Rapid	2nd Class Figure ...	200	5	Lying. 35 seconds allowed.
3	Snapshooting	" ...	200	5	Lying. Taking cover behind stones or sandbags and firing round them, with side of rifle only rested. Exposure, 6 seconds for each shot.
4	Slow	" ...	300	5	Lying. Taking cover as in 3.
5	"	" ...	300	5	Kneeling.
6	"	1st Class Figure ...	500	5	Lying.
		Total ...	...	30	

* Trained drivers will carry out Part I and Practices 8 and 10 of Part II only. The standard for qualification will be eliminated in the case of these drivers.

PART II.—STANDARD TEST.

No.	Practice.	Target.	Distance in Yards.	Rounds.	Instructions for Conduct of Practice and Standards.
7	Grouping ...	2nd Class Elementary (bull's-eye)	100	5	Lying. Qualification standard. Four shots in 12-inch ring, 5th shot not necessarily on the target. To count points, instructions in para. **443**, iii. and v., will be followed.
8	Snapshooting	2nd Class Figure ...	200	5	Lying. Taking cover as in 3. Exposure, 5 seconds for each shot. Five hits on target.
9	Rapid	" ...	200	8	Lying. One minute allowed. Chamber and magazine to be empty until the command "Rapid Fire" is given. 8 hits on the target, 6 inside the 36-inch ring.
10	Slow	1st Class Figure ...	500	5	Lying. Taking cover behind stones or sandbags representing a parapet, and firing over them. 5 hits on the target.
		Total ...	...	23	

PART III.—FIELD PRACTICES.

	Rounds.
Individual Field Practices	10
Collective Field Practices	10
Total	20*

* Or more if available from surplus rounds.

Trained soldiers who have qualified in Table A will in subsequent years fire Table B.

The practices in Part II will not be substituted for practices in Part I, and record must be kept of the practices fired by every non-commissioned officer and man, with the results of their firing.

Not more than 20 rounds, or the equivalent of 20 rounds, should be fired in one day except in the standard test or in repetition of instructional practices.

The ammunition will be allotted as follows:—

Part I	30 rounds.
Part II	23 rounds.
Part III	20 rounds.
Repetition, voluntary practices, and competitions at discretion of commanding officer	17 rounds.
Total	90 rounds.

Those who in Part II obtain 30 points will be classified as having reached the "qualification standard." No repetition of the practices in Part II will be allowed.

Those who obtain 65 points and reach the grouping standards may, as a special privilege, and with the approval of their commanding officers, in the following year fire, instead of Part II, the whole of the classification practices prescribed for the cavalry and infantry of the Regular Army, being classified according to the standards prescribed for the Regular Army. All such men will be held, whatever their classification may be, to have reached the qualification standard for that year. Any man who fails to obtain 105 points will not be permitted to fire again in the classification practices of the Regular Army until he has re-qualified in the Territorial standard test.

The whole of the surface of the 2nd Class figure target will be included in the surface for the standard of Practice 8.

Hits on the target outside the magpie circle will be recorded on the register separately from those within it, a line being ruled to separate the figures.

The Standard Tests of Tables A and B will be carried out under special arrangements to be made by general officers commanding to ensure independent supervision and marking.

R.A., R.E. (other than Field Coys.), and A.S.C.

The Royal Garrison Artillery for Defended Ports (except trained men detailed to armament machine guns), Royal Engineers (except Field Companies), and Army Service Corps, will fire annually four practices of Part I, Table A, Recruits Course, Territorial Force, which will be selected by general officers commanding-in-chief.

Permanent staff.

1. During the first two years of their appointments all the non-commissioned officers of the Permanent Staffs of the Irish Horse, Yeomanry, Field Companies Royal Engineers, and infantry will fire Part III, Table B, Regular Forces. Non-commissioned officers of the Yeomanry, Field Companies Royal Engineers, and infantry will also fire as many of the practices of Parts I and II as may be considered desirable within the limits of the ammunition allowed, in order to ensure that they are not handicapped by lack of ammunition

or opportunity for preliminary instruction prior to firing the classification practices of Part III.

2. In their third and subsequent years on the Permanent Staff, non-commissioned officers of the Irish Horse, Yeomanry, and infantry will carry out the whole of the practices of Parts I, II, and III, Table B, Regular Forces, and non-commissioned officers of Field Companies Royal Engineers will fire Parts I, II (Practices 7, 9, 12 and 14) and III.

3. *Supervision.*—Non-commissioned officers who fire the practices of Part III, Table B, Regular Forces, with a regular unit or depôt will be subject to the system of supervision which obtains in the Regular Forces, and those who fire Part III with a Territorial unit will be subject to the modified system of supervision which obtains in the Territorial Force.

4. Non-commissioned officers of the Permanent Staff should fire their annual course when attending musketry practice of the Territorial Force units to which they are posted. When, however, they are going through a refresher course with a Regular unit they may carry out their annual course of musketry with that unit.

Territorial Force Reserve.

The Territorial Force Reserve of the Yeomanry, Field Companies Royal Engineers, and infantry, will, when exercised, fire four practices, and other units two practices of Part I, Table B, Annual Course, which will be selected by general officers commanding-in-chief.

APPENDIX V.

CHANNEL ISLANDS MILITIA.

Table "A."—Recruits' Course (Engineers and Infantry).

No.	Practice.	Target.	Distance in Yards.	Rounds.	Instructions for Conduct of Practice and Standards.
		Part I.—Instructional Practices.			
1	Grouping ...	2nd Class Elementary (Bull's-eye)	100	5	Lying. All shots in 12-inch ring.
2	Application ...	2nd Class Figure ...	200	5	Lying, with arm or rifle rested. Five hits, including four within Inner (24-inch) ring.
3	" ...	" " ...	200	5	Lying. Five hits within Magpie (36-inch) ring.
4	" ...	" " ...	300	5	Lying. Five hits.
5	" ...	1st Class Figure ...	200	5	Kneeling. Four hits at least within Inner (40-inch) ring.
6	" ...	" " ...	300	5	Kneeling, with arm or rifle rested. Four hits at least within Inner (40-inch) ring.
7	" ...	" " ...	400	5	Lying. Four hits at least.
8	" ...	" " ...	500	5	Lying, with arm or rifle rested.
9	" ...	" " ...	500	5	Lying.
10	" ...	" " ...	600	5	Lying, with side of rifle only rested.
		Total rounds ...	...	50	

No.	Practice.	Target.	Distance in Yards.	Rounds.	Instructions for Conduct of Practice and Standards.
		Part II.—Classification Practices.			
11	Grouping ...	2nd Class Elementary (Bull's-eye)	100	5	Lying.
12	Application ...	1st Class Figure ...	300	5	Kneeling.
13	Rapid ...	" " ...	300	5	Lying. 40 seconds allowed.
14	Snapshooting	2nd Class Figure ...	200	5	Lying. Taking cover behind stones or sandbags representing a parapet and firing over them Exposure, 5 seconds for each shot.
15	Application ...	1st Class Figure ...	500	5	Lying.
		Total rounds ...	...	25	

Instructions.

(*a*) Recruits who fail to obtain 90 points in Part I will be classified as 3rd Class shots and repeat the whole of Part I, and will not fire Part II.

(*b*) Those who obtain 90 points in Part I but fail to reach the Grouping Standard in any of the practices in Part I, *i.e.*, 1, 2, 3, 5 and 6, will repeat once each those practices in which they have failed before proceeding to Part II.

Recruits who fire Part II will be classified upon their total scores in Part II as follows:—

Those who obtain 80 points and upwards, 1st Class shots.

Those who obtain 40 points and less than 80 points, 2nd Class shots.

Those who fail to obtain 40 points, 3rd Class shots.

104/Miscellaneous/93 Artillery Recruits will fire any four practices of Table "A."—Recruits' Course—Regular Royal Artillery. Appendix II.

Table "B."—Annual Course (Engineers and Infantry).

No.	Practice.	Target.	Distance in Yards.	Rounds.		Instructions for Conduct of Practice and Standards.
		PART I.—QUALIFYING PRACTICES.				
1	Grouping ...	2nd Class Elementary (Bull's-eye)	100	5	1st day	Lying. 4 shots in a 12-in. ring.
2	Application ...	2nd Class Figure ...	200	5		Lying. 5 hits, minimum score 14 points.
		Total rounds	...	10		

Those who do not obtain these standards will repeat once those in which they fail before proceeding to Part II.

No	Practice.	Target.	Distance in Yards.	Rounds.	Instructions for Conduct of Practice and Standards.
		Part II.—Classification Practices.			
3	Grouping ...	2nd Class Elementary (Bull's-eye)	100	5 } 1st day	Lying.
4	Snapshooting	Figure No. 3 (silhouette)	200	5 } 1st day	Lying. Taking cover behind stones or sand-bags representing a parapet and firing over them. Bayonets fixed. Exposure, 4 seconds for each shot.
5	Slow	2nd Class Figure ...	400	5 } 2nd day	Lying. Taking cover behind stones or sand-bags and firing round them, with side of rifle only rested.
6	"	" " ...	300	5 } 2nd day	Kneeling. Taking cover in a trench or behind a screen representing a wall and firing over the parapet.
7	Rapid ...	" " ...	300	15 } 2nd day	Lying. Rifle to be loaded, and four rounds in the magazine before the target

						appears. Loading from the pouch or bandolier by five rounds afterwards. 1 minute allowed. (The scoring will be as for Practice 22, Table "B," Appendix I. *See* Note 2 to para. **495**.)
8	Slow	1st Class Figure ...	500	5	3rd day	Lying.
9	Rapid	" " ...	500	5		Lying. 30 seconds allowed.
10	Slow	" " ...	600	5		Lying. Taking cover behind stones or sandbags representing a parapet and firing over them.
		Total rounds ...	...	50		

INSTRUCTIONS.

Trained soldiers will not fire Table "B" in the same year as that in which they fire Table "A."

Those who fire Part II will be classified according to the standards prescribed for the Regular Forces.

Artillery annual course will be any two practices of Table "B"—Regular Royal Artillery—Appendix II.

Annual Course—Reserves—(Engineers and Infantry)-

No.	Practice.	Target.	Distance in Yards.	Rounds.	Instructions for Conduct of Practice.
		Classification Practices.			
1	Grouping ...	2nd Class Elementary (Bull's-eye)	100	5 } 1st day	Lying.
2	Snapshooting	Figure No. 3 (silhouette)	200	5 } 1st day	Lying. Taking cover behind stones or sandbags representing a parapet and firing over them. Bayonet fixed. Exposure, 4 seconds for each shot.
3	Slow	2nd Class Figure ...	400	5 } 2nd day	Lying. Taking cover behind stones or sandbags and firing round them, with side of rifle only rested.
4	"	" " ...	300	5 } 2nd day	Kneeling. Taking cover in a trench or behind a screen representing a wall and firing over the parapet.
5	Rapid	" " ...	300	15 } 2nd day	Lying. Rifle to be loaded and four rounds in the magazine before the target appears. Loading

						from the pouch or bandolier by five rounds afterwards. 1 minute allowed. The scoring will be as for Practice 22, Table "B," Appendix I. *See* Note 2 to para. **495**.
6	Slow	1st Class Figure ...	500	5	3rd day	Lying.
7	Rapid	" " ...	500	5		Lying. 30 seconds allowed.
8	Slow	" " ...	600	5		Lying. Taking cover behind stones or sandbags representing a parapet and firing over them.
		Total rounds ...	...	50		

INSTRUCTIONS.

Reserves firing the above will be classified according to the standards prescribed for the Regular F. rces.

Reserves who come up for two days' training only will fire such practices as the G.O.C. may select from above table.

APPENDIX VI.

OFFICERS TRAINING CORPS.

Modifications to Territorial Force musketry course.

Cadets of the Officers Training Corps will fire Tables "A" and "B" as laid down for the Territorial Force, with the following modifications:—

i. Ammunition will be drawn in accordance with the scales laid down in para. 55, Equipment Regulations, Part 2, Section XVII.

ii. *Table "A."*—The 20 surplus rounds will be used for repetition and voluntary practice.

Table "B."—The 17 surplus rounds will be used for repetition and voluntary practice.

Part III will not be fired.

iii. Snapshooting, as laid down in Practices 3 and 8 of Table "B," will be substituted for rapid firing in Practices 2 and 9 of that table in cases where the latter is impracticable. The size of the target or length of exposure might be reduced in order to make the practice slightly more advanced.

iv. Cadets will be recorded as "qualified" or "failed to qualify" as the case may be, according to whether they reach or not the standards prescribed for the Territorial Force. (*See* Appendix IV.)

v. Cadets of the Senior Division who have qualified in Table "A" in the Junior Division, will not be required to re-qualify in that table.

To face page 291.

PLATE XXXIV.

HAND GRENADE (MARK I).

Scale ½.

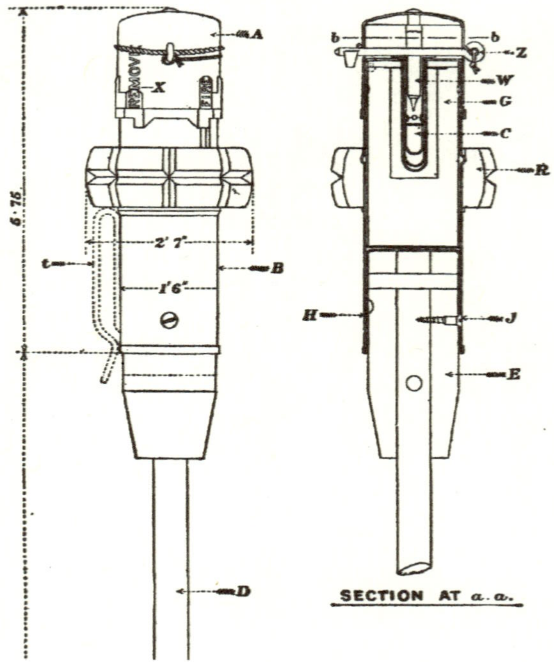

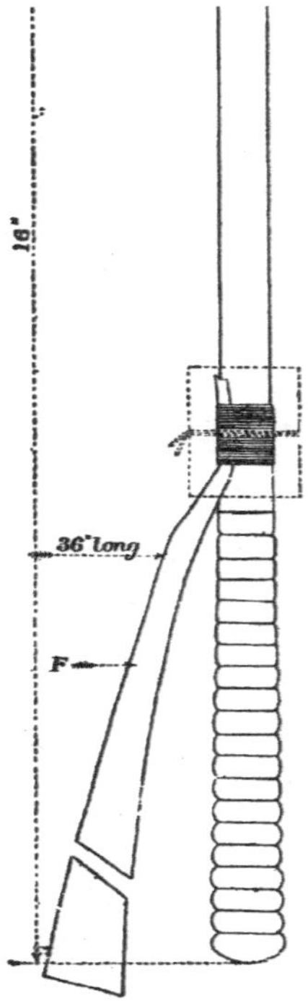

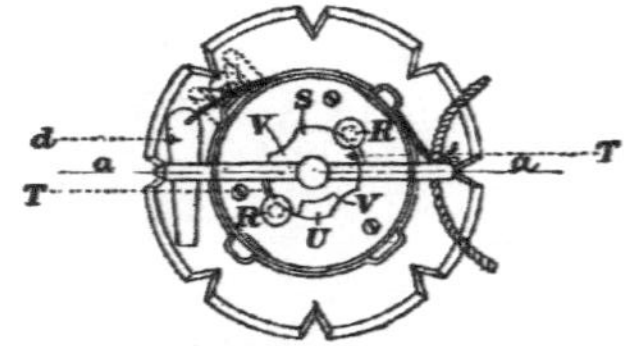

SECTION AT b.b.
GRENADE SET IN "FIRE" POSITION.

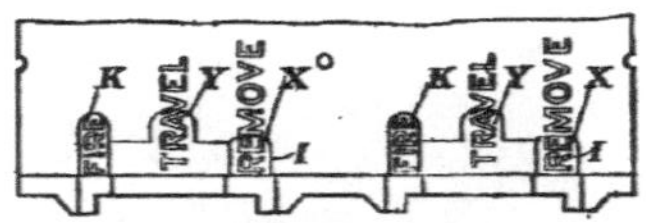

DEVELOPMENT OF CAP.

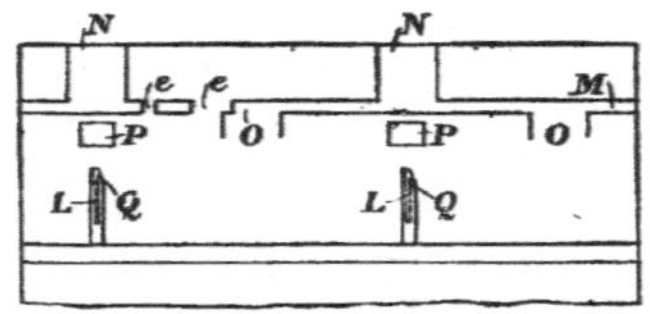

DEVELOPMENT OF GROOVE ON BODY.

APPENDIX VII.

HAND GRENADE (MARK I).

Section 1.—*General description.*

1. The grenade consists of the following principal parts:—Cap A, body B, detonator C, cane handle D, wood block E, tail F, charge G, and cast iron ring R.

2. The body B of the grenade carries the lyddite charge G; the wood block E is put into the recess in the cup H, and the cup, wood block and body are then firmly secured together by means of the three brass screws J. Attached to the wood block E is the cane handle D. to the end of which is securely bound the tail F, the cane handle D being for the purpose of throwing the grenade, and the tail F to steady it in flight and to assist to make it travel and fall point foremost.

3. The upper part of the body has a groove M formed in it for the purpose of securing the cap A in position. The groove M is provided with four leads into it, two N.N., to allow of the insertion and removal of the cap, and two O.O., to allow the cap to move forward upon the grenade striking the ground or other obstacle. Two projections, e.e., are made in the groove M for the indent X in the cap A to jump when the cap enters or leaves the travel position. One projection is to be made long enough to carry the indent into the "Fire" position. The object of these two projections is to give a definite indication of when the cap is in the "Travel" and "Fire" positions. Two indicating

knobs P.P. are secured to the body, and two stop pins Q.Q. are fixed below the indicating knobs P.P., preventing the cap A being pushed down too far (except when turned into the " Fire " position—*see* later) if by any accident the safety pin had been removed or displaced. Fixed to the top of the body are two holding studs R.R. to secure the detonator C when in position. The body has also painted on it, in red, two arrows L.L. for the purpose of indicating positions of cap A as to the removing, travel or firing positions.

4. The detonator C is formed with a flange s on which are two lugs T.T. for the purpose of turning the detonator when in position, so as to secure it under the heads of the holding studs R.R. On the face of the flange S is fixed a brass plate spring U, for locking the detonator into position. The two grooves V.V. in the flange S of the detonator C, are to allow the flange S to pass the holding studs R.R. during insertion or removal of the detonator.

5. The cap A carries a steel needle W for firing the detonator. Two small indents X.X. are formed on the cap to engage with the groove M on the body B. The raised lips I.I. are to allow the cap A to clear the indicating knobs P.P. when the cap is being placed or removed from the body of the grenade.

6. Two raised lips K.K. are to allow the cap A to move forward when the cap is turned into its firing position, the lips K.K. being raised sufficiently to clear the stop pins Q.Q., this only being possible when the cap is in the firing position. The raised portions Y.Y. are for the indicating knobs to engage with when the cap is turned to the travel position and thus give a further indication when the cap A is in this position. The cap is also fitted with a safety pin Z, which passes through

the needle and the cap and prevents the cap moving forward while the pin is in position. The pin Z is secured by a whip-cord becket passed over the cap A and is also further secured by a thin leather strip d passing through a slot at one end, it being necessary to remove both these safeguards before the pin Z can be withdrawn. The safety pin Z is also passed through the cap A in such a position that if by any mis-chance the detonator C was not properly secured after being placed in position, the act of placing the cap A on and turning it to the left into the firing position causes the pin Z to engage with the two lugs T.T. on the flange S of the detonator, and automatically locks the detonator under the heads of the holding studs R.R.

7. The hook t fixed to the body of the grenade is for attaching the latter to the soldier's belt. The grenade with the stick downward is hung on to the belt by the hook.

Section 2.—To prepare the grenade for use.

8. Turn the cap A on to the body B to the right until the indicating knobs P.P. are in the raised lips I.I. formed in the cap A. This can be seen by means of the arrows L.L. painted on the body B being opposite the words "remove" on the cap A. Then pull off the cap.

9. Place the detonator C in the recess for it. See that the two grooves V.V. in the flange S coincide with the two studs R.R., then press down the detonator into position. When the flange S is home, turn the detonator C to the left, passing the flange under the heads of the studs R.R. and continue turning until the brass plate spring U is released, thus locking the detonator.

10. Replace the cap A with the raised lips I.I. over the indicating knobs P.P. and push down into position. After the cap A has been put on, it must be turned one-eighth of a turn to the left, thus bringing the indicating knobs P.P. into the raised positions Y.Y. of the cap A. This is done by pointing the indicating arrows L.L. to "travel" on the cap.

11. The grenade is intended to be carried with the raised portions Y.Y. always over the indicating arrows L.L., *i.e.*, in the travel position whether the detonator C is in position or not.

Section 3.—To throw the grenade.

12. The tail is unwound and allowed to hang loose at full length.

13. The cap is turned from the "travel" to the "fire" position.

14. The safety pin is withdrawn.

15. The grenade is thrown by means of the cane D. The latter is grasped between the end furthest from the grenade itself and the attached point of the tail, *i.e.*, on the grooved portion. The grenade is thrown in the required direction either under or over hand, care being taken that the tail cannot entangle itself with the thrower or with any object near him.

16. When throwing, the following points should be remembered:—

i. The grenade should be thrown well upwards at not less than an angle of about 35 degrees. This, besides assisting in increasing the range to which the grenade can be thrown, renders its action more absolutely certain by causing it to strike the ground nearly vertically. This is especially important when throwing with a following wind.

ii. Any obstacle lying between the thrower and the objective must be cleared, as the grenade will almost certainly act on anything it strikes during any part of its flight.

17. Should the hand grenade not be used the cap is to be turned back to "travel" from "fire," the safety pin (which must be retained) is to be replaced in position, care being taken that the pin passes through the cap, and is secured by passing the whipcord becket over the cap, and by replacing the leather strip d through the slot in the end of the safety pin Z, and the tail rolled up and secured.

Section 4.—Instructions for use.

18. The circumstances under which hand grenades should be employed in war are detailed in "Field Service Regulations, Part I." In order to ensure an adequate supply of men capable of instructing troops in the use of the grenade, when required, a sufficient number of men will be trained in each Field Company, Royal Engineers, to ensure that six efficient instructors per company are always available.

INDEX.

A

	PARA.
Accidents, precautions against, orders to be observed...	480
Accuracy of a rifle, shown by the figure of merit ...	126
Action, cleaning of the	103
,, cocking (Webley pistol)	80
,, N.C.Os. and men not to take to pieces	112
,, of the mechanism	35, 72
,, of the mechanism (Webley pistol)	79
Advanced exercises	285
Aim corrector, use of the	209, 216, 234, 236
Aim to be taken at the lowest part of the mark	221
Aiming, accuracy in	200 *et seq.*
,, at targets described by word of mouth, practice in	211
,, at the ground and marking down an enemy ...	210
,, common faults in	206
,, during target practice	464
,, instruction in, special target for	201
,, mark for	201
,, off for movement	222
,, off for wind	212
,, rest, the	203
,, rules for	202
,, stages of instruction in	198
,, tube, instructions for fitting the	51
,, tubes, employment of	364
,, up and down	217
,, with long range sights	205
Air, resistance of the, effect of on the bullet	149
Allowance for wind (machine guns)	638A
,, of ammunition	134

PARA.

Alteration of sights, not allowed 470
Ammunition, aiming tube, extra may be drawn on payment... 368
,, allotment of, trained soldiers 421
,, allowance of 134
,, ,, to G.O.C. for inspection ... 137
,, ,, for Regulars, posted to Special Reserve battalions ... 138
,, ball and blank, precautions against taking on the parade ground 262
,, blank, cleaning rifle after firing with ... 100
,, care of 116
,, daily record of expenditure of to be kept... 141
,, daily records of expended, to be inspected by commanding officers 141
,, defective, procedure with 144
,, ,, reports on 145
,, description of... 133
,, expenditure of in machine gun practice, record of to be kept 141
,, Government, only to be used in service rifles 143
,, on hand at end of year to be carried forward 142
,, records of expended, to be examined by Accountants 141
,, service, use of on thirty yards ranges ... 369
,, surplus, from pistol practices, disposal of 476
,, surplus, use of 434
,, unexpended and surplus, distribution of 139
Angle of the fall of the bullet, the 166
,, visual, methods of measuring 318
Annual course, use of ammunition by companies prior to the 140
,, returns, instructions for ... ... 617 *et seq.*
Application ... ... 386 *et seq.*
Arms and pouches, examination of before leaving the firing point 489
,, care of, instructions for 84, 106
,, description of 12

	PARA.
Arms, examination of...	118
,, rules for conducting tests of	125
,, testing of	120
Army Ordnance Corps, courses for	Appendix II
,, Service Corps, courses for	Appendix II
,, Veterinary Corps, courses for	434A
Averages, computation of	424
Award of badges	596 *et seq.*

B

Badges, description of	597 and 598
,, issue of	596 *et seq.*
,, judging distance, award of	604
Bandoliers, pouches, etc., to be examined	262, 489
Bandsmen, exercise of	410
Barometric pressure and temperature	170 *et seq.*
Barrel, effect of a heated or oily	163
Battalion, to be read as regiment, where necessary	7
Bayonet, effect of the fixed when firing	158
,, firing with fixed	115
Beaten zone, the	180, 186
Best shooting company, prominence to be given to the...	432
,, shot, how determined	423
Blank ammunition, cleaning rifles after firing with	100
,, ,, use of for training the ear	308
Body, position of the, when firing prone	255
Bolt and muzzle, covering of, in dusty countries	113
,, cleaning the	103
,, to assemble the	46, 74
,, to remove the	94
,, to replace the	95
,, to strip the	45, 73
Bolt-head, not to be used for removing or replacing the striker	50
,, care of adapted	365 iii
Bolts not to be exchanged	110
Bore, method of cleaning the	99
,, to be wiped dry before commencing practice	163
,, use of water for cleaning the	101
,, when to be freed from oil	115

	PARA.
Boys, training of with the rifle	400
Breathing, management of the	392
,, restraint of while pressing the trigger ...	235
Browning, care of	111
Bugle sounds, when used for communication	502
Bullet, angle of the fall of the	166
,, effect of "drift" on the course of the	157
,, effect of "jump" on the course of the	155
,, effect of the force of gravity on the	149
,, effect of the resistance of the air on the	149
,, forces acting on the	149
Bull's-eye, size of on 2nd Class elementary target ...	495
,, targets, iron, method of signalling for ...	500
Butt duty, duties of officer on	485
,, officer on to be assisted by a non-commissioned officer	486
Butt memorandum, the	630
Butts and firing point, signalling between	490 *et seq.*
,, collection of metal from the	611

C

Care of ammunition	116
,, arms, instructions for	84, 106
Cartridge cases, empty, return of	137
Cartridges, dummy, use of	261
,, rifles, etc., to be examined before loading on the parade ground...	262
Casuals, exercise of	413
Cavalry, courses for	Appendix I
,, charge, effective fire against	161
Cease fire, how denoted	284
,, when used in pistol exercises	295 iii
Championships	576
Channel Islands Militia, courses for	Appendix V
Charger-loading Magazine Lee-Enfield Rifle, Mark I*, the	53 *et seq.*
Chargers, empty, return of	137
Classification, as shots, of officers and serjeants inefficient in judging distance ...	329
,, practices, and conditions of classification	427 *et seq.*, 619, xiii

PARA.
Cleaning of rifles after firing 98
" " " before firing 97
" " rusty rifles 96
" the bore, method of 99
" " rifle, general instructions for ... ... 93 et seq.
" " ·22-inch rim-fire short rifle... 105
" thorough, necessity for after using blank ... 100
Clock, dial of, as a reference to targets 279
Collective exercises, preliminary, object of 276
" field practices, see "Practices."
" fire, definition and uses of 270, 271
Combined sights not to be employed by bodies of less than two platoons 196
" " uses of 195
Command, words of, when used 281
" " " by instructors 227
Commanders, squadron and company, responsibility of for musketry efficiency 10
" to ensure that the spirit of these Regulations is observed 10
Commanding officers to report defects in arms and ammunition 117
Commands, detail of, when dispensed with 291
" during firing 288
" for fire in two ranks ... ... 293 i
Communication, maintenance of, in the absence of telephones... 502
Company, best shooting, prominence to be given to ... 432
" to be read as squadron or battery, where necessary 7
" exercises ... ... 568 et seq.
" officers and sergeants to be instructed in repairing faults met with in the field ... 118
Competitions, conduct of ... ... 572 et seq.
" surplus ammunition not to be used in practice for 139
Conditions of qualification 425, 426
Cone of fire, the 177
Conference to be held at the conclusion of a collective field practice 563

PARA.
Corrector, aim, use of 209, 216, 234, 236
Correspondence with the School of Musketry 11
Course, annual, machine gun 640
" machine gun, special instructions for ... ... 643 *et seq.*
" Recruits', Regular Forces, Cavalry, R.E., and Infantry... ... ... 401 *et seq.*
Courses, A.V.C. 434A
" cavalry, R. E., and infantry ... ... Appendix I
" Channel Islands Militia ... ... Appendix V
" Irish Horse ... ... Appendix IV
" machine gun ... ... 643 *et seq.*
" of instruction at the School of Musketry ... 658
" Officers Training Corps cadets .. .. Appendix VI
" R.A., A.S.C., and A.O.C. ... ... Appendix II
" Special Reserve ... ... Appendix III
" Territorial Force ... ... Appendix IV
" trained soldiers', special instructions for ... 415
" " " Regular Forces, cavalry, R.E., and infantry ... 408 *et seq.*
Cover for use during firing exercises to be provided ... 257
" firing from behind 446
" use of ... ... 257 *et seq.*
Criticism, suspension of during progress of practices ... 550
Cut-off and safety catch, use of 264
" not to be used to enable the rifle to be used as a single loader 264

D

Daily cleaning of rifles 96
Dangerous space, extent of 164
Defective ammunition, procedure with 144
" " reports on 145
Defects in arms and ammunition, to be reported by commanding officers 117
Definitions 147
Definition of recruit 9
" trained soldier... 9
Description of ammunition 133
" " arms 12
" " badges 597
" points for fire direction 279

PARA

Detached men, exercise of 414
Diaries ... ... 631 *et seq.*
Discipline, fire 267
Dispersion of individual and collective fire 177
Distance at which individual fire should be opened ... 268
,, judging, *see* "Judging."
Drift, amount of deflection due to 156
,, cause of 157
,, effect of on the course of the bullet 156
Dress, order of, for field practices 518
,, ,, range practices 436
Dummy cartridges, use of 261
Duties of officer on butt duty 485, 488
,, ,, superintending at the firing point on a gallery range 483
,, supervising and marking, when performed by units other than that which is firing ... 477 i

E

Eastern Command, amalgamation of R.E. units in for award of best shooting company badges 598
Effect of a heated or oily barrel 163
,, resting the rifle 162
,, the fixed bayonet 158
Efficiency, responsibility of commanders for 10
Elementary training, tests of 296
,, ,, ,, ,, machine gun 642
Elevation, how obtained 152
Emery powder, or similar material, not to be used for cleaning the rifle... 92
Empty cartridge cases, return of 137
,, chargers, return of 137
Error, triangle of, method of recording 207
Examination of arms 118
,, ,, and pouches before leaving the firing point, responsibility for 489
,, ,, bandoliers, pouches, etc., by the instructor 262

PARA.

Exemptions from musketry 412
Exercise, pistol, *see* "Pistol."
Exercises, advanced 285
,, company ... ... 568 *et seq.*
,, muscle, to be performed daily 266
,, preliminary 285
,, section and half-company ... ... 548 *et seq.*
Expenditure of ammunition, records of the 141
Extreme range, *see* "Range Table," Section 27.
Eye, distance of from the backsight 251

F

Faults, common, in aiming 206
,, complete analysis of, to be made 384
Field glasses, by whom carried 469
,, ,, practice in the use of 280
,, ,, use of 309, 533
Field practices, *see* "Practices."
Figure of Merit, computing the 126
Fingers, method of using for measuring lateral distance of targets 279
Fire, collective, comparison of results of 567
,, ,, definition and uses of 270, 271
,, direction, expenditure of rounds for 535
,, discipline 267
,, effect, ground in relation to 185
,, ,, standard tests of ... ... 555 *et seq.*
,, dispersion of 177
,, effective, against a cavalry charge 161
,, grazing 189
,, individual, definition of 268
,, ,, distance at which to be opened ... 268
,, ,, when controlled... 272
,, on the executive word, squad will halt, take cover and deliver fire 290
,, range finding by observation of 330
,, rapid, rate of 274
,, rates of 273, 274
,, rifle, theory of ... ... 146 *et seq.*
,, searching 192

	PARA.
Fire slow, rate of	274
" the beaten zone of	180
" the cone of	177
" the effective zone of	181
" in two ranks, when used	293
Firing, action when suspended	463
" at night	197
" cleaning rifles after	98
" " before	97
" commands during	288
" effect of the fixed bayonet on	158
" kneeling	256 ii
" men not engaged in to receive instructions in use of the eyes, etc.	467
" point and butts, signalling between	490
" " instruction given on...	451
" " men not actually firing to remain in rear of	465
" points, supervision of by one officer	482
" prone, position of the body	255
" " to adjust sights	254 iv
" " aim and fire	254 v
" " load	254 ii
" " lie down	254 i
" " unload	254 iii
" " when adopted	253
" rate of...	268
" rest, employment of the	231
" sitting...	256 i
" standing	240, 515
" " adjustment of wind-gauge	246 *et seq.*
" " to adjust the backsight	244
" " " long range sights ...	248
" " aim and fire	249
" " load	242
" " lower the backsight	245
" " unload	243
" up and down hill	169
" wild, to be stopped	283
" with fixed bayonet	115

PARA.

Flags, danger, use of 480
Flannelette, regulation, only to be used 88
Forces acting on the bullet 149
Foresight, method of taking correct amount of the ... 206
Fouling, removal of 102

G

Gauze, wire, size and use of 89
Glasses, field, by whom carried 469
,, ,, practice in the use of 280, 309
Government ammunition only to be used in service rifles 143
Gravity, effect of on the bullet 149
Grenade, hand, *see* "Hand."
Grit to be removed from gauze and pull through ... 89
Ground, aiming at the 210
,, hire of, for executing field practices 507
,, in relation to fire effect 185
,, training in recognising the features of the ... 309
Group, definition of 378
Grouping practices, special instructions for 443
Guns, machine, *see* "Machine."
Gymnastics, use of for developing will power 383

H

Hand grenade (Mark 1) ... ...Appendix VII
,, ,, instructions for use of the... Appendix VII, Sec. 4
,, ,, men trained in the use of the Appendix VII, Sec. 4
,, ,, to prepare for use ... ... Appendix VII, Sec. 2
,, ,, ,, throw the ... ... Appendix VII, Sec. 3
Heated barrel, effect of 163
Hill, firing up and down 169
Hits, in observation practices, signalling and value of... 498
,, ,, snapshooting practices, signalling and value of 496
,, method of signalling 495
,, percentage of to rounds, an index to steadiness ... 565
,, value of 495
,, ,, on iron bulls-eye targets 500

I

	PARA.
Inaccurate rifle, report on	124
India, commencement of musketry year in	8
,, reports and returns from	620 viii
Independent supervision	477
Individual fire, definition of	268
Infantry, courses for	Appendix I
Inspection of records of ammunition expended by commanding officer	141
,, of rifles	115
,, of tubes which are private property	365 v
,, tests	298
Instruction in aiming, stages of	198
Instructions for cleaning the rifle, general	93 *et seq.*
,, ,, rendering annual returns	617 *et seq.*
,, ,, the care of arms	84, 106
Instructors, rules for	227
Instruments, range finding, tests of	336 *et seq.*
Irish Horse, courses for	Appendix III

J

Jamb, allowance for, during timed practice	453
Judging distance, additional practice in, for all recorded as inefficient...	328
,, ,, badges for	04
,, ,, classification, as shots, of officers and sergeants inefficient in	329
,, ,, efficiency in, how determined... ...	326, 327
,, ,, mean error in	304, 305
,, ,, ,, percentage of error to be recorded	325 vi
,, ,, methods of	312
,, ,, necessity for practice in, by officers and non-commissioned officers ...	303
,, ,, over-estimation of objects, when made	315 i
,, ,, tests in	322 *et seq.*
,, ,, ,, exemptions from	325
,, ,, time allowance for	318, 325 iii

PARA.

Judging distance, to be carried out during an oversea campaign 305
,, ,, training of recruits in... ... ... 314 *et seq.*
,, ,, under-estimation of objects, when made ... ... 315 ii
,, ,, use of small scale maps as a guide for 319
,, ,, when combined with visual training 311
Jump, cause and effect of 155
,, effect of, on course of the bullet 155
,, lateral, how compensated for 156

K

Kneeling, firing when ... ... 256 ii

L

Landscape targets, use of 280
Left shoulder, shooting from the 345
Light 175
" Load," the command when used 242
Loading, in range practices... 456
,, on the parade ground, precautions to be taken before 262
,, practice in rapid 260

M

Machine gun, course of 648
,, ,, ,, special instructions for ... ... 648 *et seq.*
,, ,, ,, Special Reserve 652
,, ,, ,, Territorial Force 653
,, ,, courses at the School of Musketry ... 658
,, ,, detachments, exercise of 410
,, ,, practice, ammunition expended in, record of to be kept 141
,, ,, returns 636
Machine guns, general instructions for ... ... 637 *et seq.*
,, ,, records of grouping and traversing fire... 639
Machine gunners, 1st class, number of 641
,, ,, ,, training in range finding of 340
Magazine, care of the 109
,, charging of for rapid practices 450

PARA.
Magazine platform, to assemble the 48
,, ,, to remove the... 47
,, rounds carried in ordinary circumstances in the 242
Magazines, capacity of 61
,, for miniature cartridge rifles... 360
Mainspring, not to remain compressed 107
Maps, use of for judging distance 319
Mark for aiming 201
Mechanism, action of the 35, 72
,, ,, ,, (Webley pistol) 79
Meetings, rifle, conduct of ... ... 572 *et seq.*
Merit, figure of, computing the 126
Metal, collection of 611
,, disposal of 610
,, funds, instructions regarding ... ... 609 *et seq.*
,, ,, surplus, disposal of 614
Military shot, definition of best 451
,, vocabulary 310
Miniature cartridge practice ... ... 354 *et seq.*
,, ,, ,, rifles for 357
,, ,, range, care of arms for use on ... 365 iv
,, ,, ranges, rules for practice on ... 367
Mirrors, officers and N.C.Os. to be instructed in the use of 501
,, reflecting, code of signals for use with ... 491
Missfires, cause of 114
Movement, aiming off for 222
Muscle exercises 266
Musketry, exemptions from 412
,, not to be separated from tactical study and manœuvre training 2
,, Regulations, to whom applicable 6
,, School of, *see* "School."
,, to be associated in theory and practice with manœuvre 2
,, training, purpose of 1
,, ,, scope of 3
,, year, commencement of 8
Muzzle and bolt, covering of in dusty countries ... 115
,, not to be plugged 113

N

	PARA.
Nickelling or metallic fouling in the bore, removal of ...	102
Night firing	197
" " rests for rifles for	197

O

Observation practices	419
Occasional shots, when fired	460
Officer on butt duty, duties of	485
" " " N.C.Os. to assist	486
" superintending at firing point, duties of the ...	483
Officers, all available to be present at instructional practices	442
" of a unit firing, to assist superintending officer during tests	477
" of dismounted units, pistol practice	472 ii
" subaltern, exercise of	409
" to supervise at the butts and firing point ...	477
" Training Corps, courses for cadets of the ...	Appendix VI
Oil, paraffin, use of	91
" Russian petroleum, use of	91
" when removed from the bore	115
Oily barrel, effect of	163
Oral tests	297
Orders, framing of, by G.O's.C. to meet local conditions	435
" regimental, publication of company averages in	424
" regimental, publication of mean percentage of error in judging distance and range firing tests in	343
" to be observed in order to guard against accidents	480
" when given in anticipation	292
Orthoptics, use of, forbidden	470

P

Paraffin, use of	91
Parts of the rifle, names of, *see* footnote, p. 29.	
Penetration, testing	555

PARA

Permanent Staff, courses for ... ... Appendix IV
Petroleum, Russian, use of 91
Pistol, action of the mechanism of the 79
,, cleaning of 83
,, cocking action 80
,, description of the ... ... 77 *et seq.*
,, exercise, continuous practice ... ... 295 ii
,, ,, to return pistol ... ... 295 i
,, ,, single practice, to fire ... ... 295 i
,, ,, ,, ,, to load ... ... 295 i
,, ,, ,, ,, to unload ... ... 295 i
,, practice ... ... 471 *et seq.*
,, ,, first and subsequent years 472
,, ,, method of conducting 474
,, ,, register for ... ... 475 iii
,, ,, with ball and blank 471, 472
Platform, magazine, to assemble the 48
,, ,, ,, remove the 47
Plug, insertion of in muzzle forbidden 113
Positions, correct firing, not to be departed from ... 391
,, firing, for field practices 514
,, ,, in two ranks 294
,, regulation, obligatory in range practices ... 455
,, ,, when varied 229
Pouches, examination of 262, 489
Powder, emery, or similar material, not to be used for cleaning rifles 92
Practice, aiming and snapping during 464
,, bore to be wiped dry before commencement of 163
,, miniature cartridge ... ... 354 *et seq.*
,, on miniature cartridge ranges, rules for ... 367
Practices, classification, and conditions of classification 427 *et seq.*
,, collective field ... ... 540 *et seq.*
,, ,, object of 542
,, ,, programme for 540
,, ,, ranges for 543
,, ,, scope of instruction 544
,, criticism to be suspended during progress of defence 550
,, ... 549
,, field, allotments of ammunition for 518

PARA.
Practices, field, criticism of, to be complete 564
" " firing positions for 514
" " general programme for 516
" " general rules for ... ... 503 *et seq.*
" " hire of ground for the execution of ... 507
" " importance of 4
" " individual conduct of ... ... 524 *et seq.*
" " individual, distances for 529
" " " when fired 5
" " method of recording ricochets in ... 520
" " number of rounds to be fired in one day 512
" " order of dress for 513
" " records of, to be produced at inspections 520
" " special instructions for ... ... 512 *et seq*
" " targets for 546
" " use of, classification ranges for 517
" fire direction ... ... 532 *et seq.*
" fired by R.E. 434
" grouping, special instruction for 443
" instructional, all available officers to be present at 442
" instructional, firing of without signalling ... 419
" " range ... ... 372 *et seq.*
" " when varied 417
" observation 419
" " signalling and value of hits ... 497, 498
" preliminary and qualifying, division of ... 377
" range, *see* "Range."
" rapid, timing in 448
" repetition 398
" slow, time limit for 445
" snap shooting 396
" " target for 497
" when duties will be performed by units other than that which is firing 477
Precautions to be taken before loading takes place on the parade grounds 262
Preliminary exercises 285
" training of recruits ... ... 343 *et seq.*
" " " trained soldiers ... ... 351 *et seq.*

PARA

Pressing the trigger, method of 232
Prizes, award of 605
,, forfeiture of 607
Prone, firing when, *see* "Firing."
Pull-off, defect in 108
,, weight of 108
Pull-through, cord of, not to rub against muzzle or chamber 88
,, how packed 87
,, to be free from grit 89
,, ,, drawn through in one motion ... 88

Q

Qualification, conditions of 425, 426
Qualifications, record of ... ... 665 *et seq.*

R

R.A.M.C., course for recruits of the 434A
Range, extreme, *see* "Range Table, Section 27."
,, miniature, care of arms and tubes for use on ... 365 iv
Ranges, classification, use of for field practices ... 517
,, less than full extent 439, 440
,, miniature cartridge, care of 366
,, non-gallery rifle, instructions for ... ... 488 *et seq.*
,, thirty yards... ... ... 369 *et seq.*
Ranging by maps, range-finders, and the eye in combination 321
,, general remarks on ... ... 301 *et seq.*
,, means of 301
,, probable error in 195
Range duties ... ... 477 *et seq.*
Range-finders, use of during company exercises ... 571
,, ,, in fire direction practices ... 538
Range-finding, training in of battalion machine gunners 340
,, by observation of fire 330
,, instruments, test of 336

PARA.
Range practices, (*see also* "Practices") ... ... 343 *et seq.*
,, general instructions for the conduct of 435 *et seq.*
,, instructional ... ... 372 *et seq.*
,, ,, to precede field practices 5
,, order of dress for 436
,, ,, firing 438
,, regulation positions obligatory in ... 455
,, special instructions for... ... 389 *et seq.* 443 *et seq.*
,, to be fired in favourable weather ... 437
Range table of the S.M.L.E. rifle 168
,, takers, numbers maintained 337
,, ,, test of ... ... 337 *et seq.*
,, warden to perform minor fatigues... 481
Rapid firing, acquired skill in to be developed 508
,, loading, practice in 260
Rates of fire 273, 274
Record, daily, of ammunition expended, to be kept ... 141
Records, general instructions regarding ... ... 616 *et seq.*
,, of ammunition expended, examination of by accountants 141
,, of results, filling of by battalions 559
Recruit, definition of... 9
Recruits, method of explaining the trajectory to ... 151
,, not to fire in cold weather 437
,, preliminary training of ... ... 343 *et seq.*
,, Course, Regular Forces, cavalry, R.E. and infantry ... ... 401 *et seq.*
Reflectors, use of 545
Registers ... ... 623 *et seq.*
Report of defects in arms and ammunition to be made by commanding officers 117
,, on an inaccurate rifle 124
Reports, general instruction regarding ... ... 616 *et seq.*
,, of defective ammunition 145
Resistance of the air, effect of on the bullet 149
Rest, aiming 203
,, firing, employment of the 231
Rests for rifles for night firing 197
Return, musketry prize 622
,, ,, transfer 621
Returns and records, destruction of 616

PARA.
Returns, annual, instructions for ... ... 617 *et seq.*
„ general instructions regarding ... ... 616 *et seq.*
„ machine gun 636
„ Territorial Force ... ... 633 *et seq.*
Revolver, all armed with to be practised in the use of ... 295
Ricochets 168
„ how discerned on the target 494
„ „ shown in records of results of field practices 520
Rifle, cleaning the, *see* "Cleaning."
„ effect of resting the 162
„ fire, practical application of ... ... 146 *et seq.*
„ „ theory of ... ... 146 *et seq.*
„ meetings, conduct of ... ... 572 *et seq.*
„ soldier's own to be used 458
„ support of 462
Rifle ranges, non-gallery, instructions for ... ... 488 *et seq.*
Rifles, rests for, for night firing 197
„ sighting of 153
Rifling 148
Rounds, not fired, to be forfeited 454
„ number of, to be fired in one day 441
„ „ „ at field practices 512
Royal Artillery, courses for ... ... Appendix II
Royal Engineers, courses for ... ... Appendix I
„ practices fired by 434
„ training of in the use of the hand grenade ... ... Appendix VII, Sec. 4
Rule for corrections for barometric pressure 172
Rules for aiming 202
„ conducting tests of arms 125
Russian petroleum, use of 91
Rusty rifles, cleaning of 96

S

Safety, general officers commanding to frame orders for 435
„ catch, procedure when devoid of the 265
„ „ and cut-off, use of 264
„ „ when applied 264, 289

PARA.

School of Musketry, appointment of sergeant instructors at 668
" " attendance at the ... ... 660 *et seq.*
" " correspondence with the ... 11
" " courses of instruction at the... 658
" " duties of the commandant of the 655
" " object of the ... ... 654 *et seq.*
" " record of qualification gained at the ... ... 665 *et seq.*
Scoring for pistol practice ... ... 475 i
Screens, use of for testing fire effect 556
Searching fire 192
Serjeant instructors, appointment of 668
Shooting, timed, high degree of proficiency required in 394
Short Magazine Lee-Enfield rifle, the ... ... 12 *et seq.*
" Rifle, ·22 inch Rim fire, Mark I, the ... ... 49 *et seq.*
Shot cutting outer edge of any ring, how counted ... 494
" military, definition of 451
" not wholly on the target, how counted 494
Shots, occasional, when fired 460
" sighting, not allowed 459
" signalling of ... ... 492 *et seq.*
Sight, line of, height of trajectory above the 168
Sights, combined, uses of 195
" " use of by small units 196
" inspection of the adjustment of 202
" long-range, aiming with 205
" rapid adjustment of 219
" to be used as issued without alteration of any kind 470
Sighting, correction of 418
" method of finding incorrect 128
" of rifles 153
" shots, use of, prohibited 418, 459
Signals, code of, for use with reflecting mirrors ... 491
Signallers, exercise of 410
Signalling between the butts and the firing point ... 490 *et seq.*
" hits, method of 495
" " in snapshooting practices 496

PARA.

Signalling, for iron bull's-eye targets, method of ... 500
,, in machine gun practices 649
,, shots ... ... 492 *et seq.*
Sitting, firing when ... ... 256 i
Sling, use of the 239
,, in range practices 457
Sloping arms 265, 289
Slow practices, time limit for 445
Small arms, examination of 118
Smoking, excessive cigarette, cause of unsteadiness ... 380
Snapping, daily practice in 238
,, during target practice 464
Snapshooting, acquired skill in to be developed ... 508
Sound, locating the enemy by, use of blank ammunition for practice in 308
Southern Command, amalgamation of R.E. units in for award of best shooting company badges 598
Space, dangerous, extent of.. 165
Special Reserve, commencement of musketry year for the 8
,, ,, courses for ... ... Appendix III
,, ,, machine gun course for the 652
,, ,, Regular Establishment, courses for ... 138
,, ,, supplementary instructions included in Regulations for 6
Squads, strength of 228
Standard figures, record of to be kept 519, 559
Standing, firing when, *see* "Firing."
Sub-target machine, practice in trigger pressing with the 236, 238
Supervision, independent 477
Suspension of fire, action on 463

T

Table "B," 2nd class standard in, men unable to reach to be reported specially 420
,, "C" 643
,, showing effect of fixed bayonet on the trajectory 159
Tables "A" and "B" ... ... Appendix I
,, ,, execution of in the same year ... 404
Tactical schemes ... ... 548 *et seq.*
,, ,, rehearsal of 552

	PARA.
Target, for pistol practice	474 ii
,, indistinct, necessity for training in picking up	509
,, special, for aiming instruction	201
Targets, allotment of...	482
,, described by word of mouth, practice in aiming at	211
,, description of	277
,, disappearing, use of	549
,, exposure of for rapid fire practices	447
,, for field practices	546
,, iron bull's-eye, method of signalling for ...	500
,, landscape, use of	280
,, ,, and figure, use of on miniature ranges	358
,, moving, aiming at	222 *et seq.*
,, observation practices	497
,, one system only of describing in battalions ...	278
Team competitions	582
Telephones, officers and N.C.Os. to be instructed in the use of	501
Temperature and barometric pressure	170
Territorial Force, commencement of musketry year for the	8
,, ,, courses for the	Appendix IV
,, ,, machine gun course for the	653
,, ,, returns for the	633 *et seq.*
,, ,, supplementary instructions included in Regulations for...	6
Test of range finding instruments	336 *et seq.*
,, ,, takers	338 *et seq.*
Testing arms	120
Tests, comparative, of fire effect and vulnerability ...	560 *et seq.*
,, inspection	298
,, ,, officers and men from units other than that firing to perform duties at ...	477
,, method of conducting	297
,, of elementary training	296
,, ,, ,, machine gun	642
,, oral	297
,, standard	299

PARA.
Tests, standard, of fire effect ... ... 555 *et seq.*
„ „ results of to be recorded 296
„ in judging distance ... ... 322 *et seq.*
„ „ „ exemptions from 325
„ of arms, rules for conducting 125
Theory of rifle fire, importance of a knowledge of the ... 146
Third class shots, employment of 433
„ „ „ not to fire when trying for observation 331
Thirty yards ranges ... ... 369 *et seq.*
Training, preliminary, of recruits ... ... 343 *et seq.*
„ „ trained soldiers ... ... 351 *et seq.*
„ „ standard tests in, object of ... 296
„ visual, *see* "Visual."
Trained soldier, definition of 9
„ soldiers' courses 408
„ „ „ special instructions for ... 415
„ soldiers, preliminary training of... ... ... 351 *et seq.*
Trajectory, method of explaining the, to recruits ... 151
„ table showing effect of the fixed bayonet on 160
„ „ „ height of in feet above the line of sight 168
Triangle of error, method of recording the 207
Trigger pressing 232
Tube, aiming, instructions for fitting the 51
Tubes, aiming, employment of 364

U

Uniform, coat or waterproof sheet to protect the ... 462
Unsteadiness, causes of 380
"Unserviceable" rifles not necessarily "inaccurate" ... 182A

V

Visual angle, methods of measuring 318
„ training, exercises in 307
„ „ general remarks on ... ... 301 *et seq.*
„ „ subjects included in 308
„ „ when combined with judging distance... 311
Vocabulary, military... 310
Volleys, less effective than bursts of independent fire... 283
Vulnerability, comparative tests of ... ... 560 *et seq.*

W

	PARA.
Water, use of, for cleaning the bore	101
Waterproof sheets	462
Weather favourable, range practices to be fired in ...	437
Weight of the pull-off	108
Will power, development of...	383
Wind	173
,, aiming off for	212
,, table, elaborate, use of prohibited	418
Wind-gauge, alteration of during practices	452
,, ,, minute adjustments of the	387
,, ,, use of the	212
,, ,, ,, for miniature cartridge practice ...	359
Wire gauze, use of	89
Words of command, *see* "Commands."	

Y

Year, ammunition on hand at end of, disposal of ...	142
,, execution of Tables "A" and "B" in the same	404
,, first, pistol practice in	472 i
,, musketry, commencement of	8
Years, pistol practice in subsequent	472 ii

Z

Zone, beaten, of machine gun fire	638
,, ,, on level ground	186
,, ,, ,, rising ground	187
,, ,, the	180
,, of effective fire, the	181